MODULES AND THE STRUCTURE OF RINGS

PURE AND APPLIED MATHEMATICS

A Program of Monographs, Textbooks, and Lecture Notes

MONOGRAPHS AND TEXTBOOKS IN
PURE AND APPLIED MATHEMATICS

30. *J. S. Golan,* Localization of Noncommutative Rings (1975)
31. *G. Klambauer,* Mathematical Analysis (1975)
32. *M. K. Agoston,* Algebraic Topology: A First Course (1976)
33. *K. R. Goodearl,* Ring Theory: Nonsingular Rings and Modules (1976)
34. *L. E. Mansfield,* Linear Algebra with Geometric Applications: Selected Topics (1976)
35. *N. J. Pullman,* Matrix Theory and Its Applications (1976)
36. *B. R. McDonald,* Geometric Algebra Over Local Rings (1976)
37. *C. W. Groetsch,* Generalized Inverses of Linear Operators: Representation and Approximation (1977)
38. *J. E. Kuczkowski and J. L. Gersting,* Abstract Algebra: A First Look (1977)
39. *C. O. Christenson and W. L. Voxman,* Aspects of Topology (1977)
40. *M. Nagata,* Field Theory (1977)
41. *R. L. Long,* Algebraic Number Theory (1977)
42. *W. F. Pfeffer,* Integrals and Measures (1977)
43. *R. L. Wheeden and A. Zygmund,* Measure and Integral: An Introduction to Real Analysis (1977)
44. *J. H. Curtiss,* Introduction to Functions of a Complex Variable (1978)
45. *K. Hrbacek and T. Jech,* Introduction to Set Theory (1978)
46. *W. S. Massey,* Homology and Cohomology Theory (1978)
47. *M. Marcus,* Introduction to Modern Algebra (1978)
48. *E. C. Young,* Vector and Tensor Analysis (1978)
49. *S. B. Nadler, Jr.,* Hyperspaces of Sets (1978)
50. *S. K. Segal,* Topics in Group Rings (1978)
51. *A. C. M. van Rooij,* Non-Archimedean Functional Analysis (1978)
52. *L. Corwin and R. Szczarba,* Calculus in Vector Spaces (1979)
53. *C. Sadosky,* Interpolation of Operators and Singular Integrals: An Introduction to Harmonic Analysis (1979)
54. *J. Cronin,* Differential Equations: Introduction and Quantitative Theory (1980)
55. *C. W. Groetsch,* Elements of Applicable Functional Analysis (1980)
56. *I. Vaisman,* Foundations of Three-Dimensional Euclidean Geometry (1980)
57. *H. I. Freedman,* Deterministic Mathematical Models in Population Ecology (1980)
58. *S. B. Chae,* Lebesgue Integration (1980)
59. *C. S. Rees, S. M. Shah, and C. V. Stanojević,* Theory and Applications of Fourier Analysis (1981)
60. *L. Nachbin,* Introduction to Functional Analysis: Banach Spaces and Differential Calculus (R. M. Aron, translator) (1981)
61. *G. Orzech and M. Orzech,* Plane Algebraic Curves: An Introduction Via Valuations (1981)
62. *R. Johnsonbaugh and W. E. Pfaffenberger,* Foundations of Mathematical Analysis (1981)
63. *W. L. Voxman and R. H. Goetschel,* Advanced Calculus: An Introduction to Modern Analysis (1981)
64. *L. J. Corwin and R. H. Szcarba,* Multivariable Calculus (1982)
65. *V. I. Istrătescu,* Introduction to Linear Operator Theory (1981)
66. *R. D. Järvinen,* Finite and Infinite Dimensional Linear Spaces: A Comparative Study in Algebraic and Analytic Settings (1981)

67. *J. K. Beem and P. E. Ehrlich*, Global Lorentzian Geometry (1981)
68. *D. L. Armacost*, The Structure of Locally Compact Abelian Groups (1981)
69. *J. W. Brewer and M. K. Smith, eds.*, Emmy Noether: A Tribute to Her Life and Work (1981)
70. *K. H. Kim*, Boolean Matrix Theory and Applications (1982)
71. *T. W. Wieting*, The Mathematical Theory of Chromatic Plane Ornaments (1982)
72. *D. B. Gauld*, Differential Topology: An Introduction (1982)
73. *R. L. Faber*, Foundations of Euclidean and Non-Euclidean Geometry (1983)
74. *M. Carmeli*, Statistical Theory and Random Matrices (1983)
75. *J. H. Carruth, J. A. Hildebrant, and R. J. Koch*, The Theory of Topological Semigroups (1983)
76. *R. L. Faber*, Differential Geometry and Relativity Theory: An Introduction (1983)
77. *S. Barnett*, Polynomials and Linear Control Systems (1983)
78. *G. Karpilovsky*, Commutative Group Algebras (1983)
79. *F. Van Oystaeyen and A. Verschoren*, Relative Invariants of Rings: The Commutative Theory (1983)
80. *I. Vaisman*, A First Course in Differential Geometry (1984)
81. *G. W. Swan*, Applications of Optimal Control Theory in Biomedicine (1984)
82. *T. Petrie and J. D. Randall*, Transformation Groups on Manifolds (1984)
83. *K. Goebel and S. Reich*, Uniform Convexity, Hyperbolic Geometry, and Nonexpansive Mappings (1984)
84. *T. Albu and C. Năstăsescu*, Relative Finiteness in Module Theory (1984)
85. *K. Hrbacek and T. Jech*, Introduction to Set Theory, Second Edition, Revised and Expanded (1984)
86. *F. Van Oystaeyen and A. Verschoren*, Relative Invariants of Rings: The Noncommutative Theory (1984)
87. *B. R. McDonald*, Linear Algebra Over Commutative Rings (1984)
88. *M. Namba*, Geometry of Projective Algebraic Curves (1984)
89. *G. F. Webb*, Theory of Nonlinear Age-Dependent Population Dynamics (1985)
90. *M. R. Bremner, R. V. Moody, and J. Patera*, Tables of Dominant Weight Multiplicities for Representations of Simple Lie Algebras (1985)
91. *A. E. Fekete*, Real Linear Algebra (1985)
92. *S. B. Chae*, Holomorphy and Calculus in Normed Spaces (1985)
93. *A. J. Jerri*, Introduction to Integral Equations with Applications (1985)
94. *G. Karpilovsky*, Projective Representations of Finite Groups (1985)
95. *L. Narici and E. Beckenstein*, Topological Vector Spaces (1985)
96. *J. Weeks*, The Shape of Space: How to Visualize Surfaces and Three-Dimensional Manifolds (1985)
97. *P. R. Gribik and K. O. Kortanek*, Extremal Methods of Operations Research (1985)
98. *J.-A. Chao and W. A. Woyczynski, eds.*, Probability Theory and Harmonic Analysis (1986)
99. *G. D. Crown, M. H. Fenrick, and R. J. Valenza*, Abstract Algebra (1986)
100. *J. H. Carruth, J. A. Hildebrant, and R. J. Koch*, The Theory of Topological Semigroups, Volume 2 (1986)

Other Volumes in Preparation

MODULES AND THE STRUCTURE OF RINGS

A Primer

Jonathan S. Golan
University of Haifa
Haifa, Israel

Tom Head
State University of New York at Binghamton
Binghamton, New York

Marcel Dekker, Inc.   **New York • Basel • Hong Kong**

Library of Congress Cataloging--in--Publication Data

Golan, Jonathan S.
    Modules and the structure of rings: a primer/Jonathan S. Golan,
Tom Head.
        p.   cm. -- -- (Monographs and textbooks in pure and applied
mathematics; 147)
    Includes bibliographical references (p.   ) and index.
    ISBN 0-8247-8555-X (acid-free paper)
    1. Modules (Algebra) 2. Rings (Algebra) I. Head, Tom.
II. Title. III. Series.
QA247.G645   1991
512'.4-- --dc20
                                                      91-8369
                                                          CIP

This book is printed on acid-free paper.

MARCEL DEKKER, INC.
270 Madison Avenue, New York, New York 10016

Current printing (last digit):
10 9 8 7 6 5 4 3 2 1

PRINTED IN THE UNITED STATES OF AMERICA

To our children,

Aharon Yosef,
Elitsur Yaakov, and
Yael Dvorah Golan

Tawn Elise,
Roark Alan,
Michael Reuben, and
Hadassah Joann Head

Preface

This book is an expanded version of the second author's volume "Modules: a Primer of Structure Theorems" [1974], revised and updated by the first author. Like its predecessor––now out of print––it is designed for use as a text for an upper–division or graduate–level one–semester introductory course in ring theory for students who have had some experience with linear algebra and at least one solid semester of abstract algebra, which can then serve as a stepping–stone for a more advanced course in homological algebra or module theory.

No previous acquaintance with modules is assumed, though we presume the student has some elementary knowledge of the theory of rings. Notions from category theory have been avoided. Infinite ordinal numbers and transfinite constructions have been used in only three sections: §4.3, §7.3, and §12.2. The Axiom of Choice is needed throughout, for the most part in the form explicitly presented in §0.11.

Jonathan S. Golan

Tom Head

Contents

Part One: Projectivity

Part Two: Injectivity

Contents

Introduction

WHAT THIS BOOK IS ABOUT

The structure of vector spaces over a field is easily described: an axiom of mathematics equivalent to the Axiom of Choice states that any vector space over a field has a basis. Thus, if we assume the Axiom of Choice (and we *will* always assume it, for why be poor by volition), we readily observe that any vector space over a field is isomorphic to a coproduct of copies of the field——and we are essentially done. If, however, we want to generalize the notion of a vector space by allowing our scalars to come from an arbitrary ring, we enter the realm of module theory and here things become much more complicated and also much more interesting.

This volume will introduce you to how algebraists deal with the problem of the structure of modules over rings and how they make use of these structures to classify rings. It is by no means complete——the structures and techniques in use are far too comprehensive to be even properly surveyed in an introductory text——but will hopefully give you a feeling for the type of work being done and the way mathematicians go about it.

SOMETHING ABOUT STYLE

Since this book is conceived as a guidebook, and not as an encyclopedia, we have chosen to write in a style which is rigorous but not overly formal. Deliberately, we have shied away from some of the more ponderous mathematical conventions. Instead of talking about lemmata (this is the correct plural of the word "lemma", which is derived from the Greek $\lambda\eta\mu\mu\alpha$, meaning "something taken for granted, an argument"), or propositions, we prefer to talk about observations (this harks back to the original meaning of the word "theorem", which is derived from the Greek verb $\theta\varepsilon\omega\rho\varepsilon\iota\nu$, meaning "to look at"); instead of giving proofs we present verifications. This terminology also serves to emphasize the essential inductive aspect of mathematical research, often all too hidden behind the deductive structure of published mathematics––the working mathematician observes relations among the structures under consideration and only later tries to verify these observations by constructing rigorous proofs.

Mathematicians like to cover their tracks and so mathematical works are usually written in a step–by–step logical manner, each result piled upon the others, leaving the reader to figure out for himself why and how the author chose this particular path on the way to a solution of the problems to which he has addressed himself. The result is often formal beauty but intuitive and pedagogical sterility. We have chosen, instead, to chart our intuitive course at each stage, explaining why such–and–such seems to be the most promising approach and where we expect to be at the end of the next stage. When appropriate, we do not hesitate to address you, the reader, directly.

A CAUTIONARY NOTE ON MATHEMATICAL SHORTHAND AND SLANG

Mathematical symbols and terms have exact meanings, precisely and formally defined, and the practice of mathematics involves the careful manipulation of these terms and symbols. However, formalism, carried to excess, leads to notation so cumbersome and distinctions so fine that the thread of argument is often lost in the process. Therefore working mathematicians, like everybody else, usually introduce shorthand and slang into their writing to make it easier to follow. A simple example illustrates this: if A, B, and C are nonempty sets then a function f from the cartesian product $A \times B$ to C is a rule which assigns to each ordered pair (a,b) in $A \times B$ an element $f(a,b)$ of C. In particular, if A, B, and C are all taken to be the same additive abelian group G then addition is such a function. However, our calculations would be hidden among a mass of unnecessary parentheses if we insisted on writing the sum of g and g' as $+(g,g')$ and the sum of g, g', and g" as $+(+(g,g'),g")$. We therefore introduce the shorthand notation $g + g'$ and $g + g' + g"$, knowing that in the second case there is no ambiguity since addition in G satisfies the associative law. Now suppose that we have an infinite set $\{g_i \mid i \in \Omega\}$ of elements of G. Since infinite addition is not defined, as a rule, in abelian groups, the sum $\Sigma\{g_i \mid i \in \Omega\}$ makes no sense. However, if we happen to know that only finitely–many of the elements g_i are nonzero, then we introduce this as a slang expression meaning the same as $\Sigma\{g_i \mid i \in \Omega$ and $g_i \neq 0\}$.

One has to be very careful, however, in introducing mathematical slang and shorthand. At each stage you must be absolutely sure that some essential distinction or essential aspect of your definition has not been lost in the process. Mathematicians, as we have said, do it all the time. Students had best beware until they are on sure footing.

FUNDAMENTAL CONCEPTS

0

Modules and Maps

§1 MODULES

Each ring R which we consider will be assumed to contain a multiplicative identity element, denoted by 1, which differs from its additive identity element 0. Now let R be a ring.

DEFINITION. A *left* R-*module* consists of an additive abelian group A together with a function μ from R × A to A which satisfies the following four conditions:

(1) $\mu(r, a+a') = \mu(r,a) + \mu(r,a')$,

(2) $\mu(r+r', a) = \mu(r,a) + \mu(r',a)$,

(3) $\mu(rr',a) = \mu(r, \mu(r',a))$, and

(4) $\mu(1,a) = a$,

for all elements r and r' of R and all elements a and a' of A.

By convention, the element $\mu(r,a)$ is denoted by r·a or by ra, and μ is said to provide an operation of *scalar multiplication* on the abelian group A. Using this juxtaposition notation instead of the function μ, the above four conditions become:

3

(1) $r(a + a') = ra + ra'$,
(2) $(r + r')a = ra + r'a$,
(3) $(rr')a = r(r'a)$, and
(4) $1a = a$

for all elements r and r' of R and all elements a and a'
of A.

A purist will observe that a left R–module is an ordered
pair (A,μ) in which the first member, A, is an abelian group
and the second member, μ, is a function. In referring to the
module (A,μ), it will always be sufficiently clear to mention
only the abelian group A, deleting reference to the function μ,
in the same way that we freely talk about "the abelian group
A"   and not   "the abelian group $(A,+)$".   Thus, we will
continually begin discussions with such (hopefully) unambiguous
statements as "Let   A   and   B   be left R–modules".   Packed
into such a statement are the assumptions that   R   is a ring
with multiplicative identity which differs from its additive
identity, that   A   and   B   are additive abelian groups, and that
there   are   functions μ:R × A → A   and
μ':R × B → B satisfying the four conditions stated above. As
pointed out in the introduction, without such verbal contractions,
mathematical statements would surely become too unwieldy to
be of use.

We are now ready for some examples of modules. If R is
a field then left R–modules are also called   *vector spaces*
over   R.   Elementary courses in linear algebra deal with vector
spaces, especially those over the field of real numbers, and
consequently supply an immediate collection of examples of
modules.

Every additive abelian group   A   may be regarded as a left
Z–module, where **Z** is the ring of integers, in precisely one
way: for n in **Z** and a in A, define na as follows:
(1) If $n > 0$, then na is the sum of n copies of a,
(2) If $n = 0$, then na = 0,
(3) If $n < 0$ then na is the sum of −n copies of −a.
It is elementary to verify that the four conditions specified in
the definition of a left **Z**–module are satisfied by this operation

and that this is the only operation of scalar multiplication on $\mathbb{Z}$ and A which is compatible with these conditions (see Exercise 1). Because each abelian group allows precisely one $\mathbb{Z}$-module structure (and because of some additional elementary facts to be pointed out later in this chapter) the theory of abelian groups is indistinguishable from the theory of left $\mathbb{Z}$-modules and is therefore subsumed by module theory.

Any ring R can be considered as a left R-module in the following manner: for the abelian group use R with its additive structure. For the scalar multiplication of the ring R on the abelian group (R,+), use the ring multiplication. Each of the four conditions in the definition of an R-module follows from the usual defining conditions in the definition of a ring (see Exercise 2). From the point of view developed in this book, the ring R regarded in this way as a module over itself is the starting point for the structure theory of left R-modules. Chapter 1 is devoted entirely to an examination of R as a module over itself, and the remainder of the book is an investigation of how further left R-modules can be developed from R by means of constructions described in this chapter and one additional construction introduced in Chapter 7.

The "smallest" of all left R-modules is the one having precisely one element, namely an additive identity. We will consistently denote this module by (0).

Our definition of a left R-module A above was based on a function from R × A to A. If we reverse the order of R and A in this cartesian product and make further notational changes that seem natural, we obtain the definition of a right R-module.

DEFINITION. A *right R-module* consists of an additive abelian group A together with a function $\mu{:}A \times R \to$ A, which satisfies the following conditions:

(1) $\mu(a+a', r) = \mu(a,r) + \mu(a',r)$,

(2) $\mu(a, r+r') = \mu(a,r) + \mu(a,r')$,

(3) $\mu(a, rr') = \mu(\mu(a,r),r')$, and

(4) $\mu(a,1) = a$,

where r, r' are elements of R and a, a' are elements of A.

As in the case of left R–modules, the element $\mu(a,r)$ is conventionally denoted by a·r or simply by ar. Using the juxtaposition notation instead of the function μ, the above four conditions can be written as follows:

(1) $(a + a')r = ar + a'r$,

(2) $a(r + r') = ar + ar'$,

(3) $a(rr') = (ar)r'$, and

(4) $a1 = a$.

It might appear that left and right modules differ only notationally and not mathematically. This would be true except for the presence of condition (3), which constitutes a genuine mathematical distinction: for right modules the result of applying the product of two scalars to a module element is the same as the effect of applying the *first* scalar and then the *second*; whereas, for left modules, the effect is the same as applying the *second* scalar and then the *first*. For rings which are not commutative, this may make a big difference. Therefore, when talking about modules over rings which are not necessarily commutative, we will have to be sure to indicate whether we are talking about left or right scalar multiplication.

If A is a vector space of finite dimension n over a field F and if S is the full ring of n×n matrices with entries in F, then A is a right S–module, with scalar multiplication being the usual multiplication of a vector by a matrix. Also, we have already noted that any ring can be considered as a left module over itself; by the same sort of argument, it can also be considered as a right module over itself. In §4 we will present an important method of constructing right modules from left modules and vice versa.

For the most part we will study left modules, though occasionally we will need right ones as well. Since each of our results on left modules has a natural reinterpretation as a result on right modules, we will not explicitly state the "right–handed" version of every "left–handed" result or definition we give, but will allow the reader to fill in the details when necessary.

EXERCISES

1. Show that the operation of the ring $\mathbb{Z}$ of integers on an abelian group A, as described above, indeed provides A with the structure of a left $\mathbb{Z}$-module and that, indeed, this is the only $\mathbb{Z}$-module structure possible on A.

2. Verify that the four conditions listed in the definition of a left module are satisfied by the operation of the ring R on the additive abelian group (R,+) that, as discussed above, is provided by ring multiplication. Similarly, show that R satisfies the four conditions listed in the definition of a right module.

3. Let S be a subring of a ring R (all subrings have the same additive and multiplicative identity elements). Show that any left R-module is also a left S-module. Give an example of a subring S of a ring R and of an additive abelian group A which is an S-module but not an R-module.

4. Show that the set F^n of all n-tuples of elements of a field F is, in a natural way, a right F_n-module, where F_n is the ring of all n$\times$n matrices with entries from F.

§2 SUBMODULES AND FACTOR MODULES

Let A be a left R-module. A subset B of A is a *submodule* of A if it is a subgroup of A and if rb is in B whenever r is in R and b is in B. If R is a field then the submodules of a left R-module (that is to say, a vector space over R) are precisely the familiar subspaces. If A is an additive abelian group considered, as above, as a left

$\mathbb{Z}$–module then its submodules are just its subgroups. Without this fact, we could not fully identify the theory of abelian groups with the theory of left $\mathbb{Z}$–modules. Submodules of right R–modules are, of course, defined in a similar fashion.

If I is a two–sided ideal of R and if A is a left R–module then the set IA consisting of all elements of A of the form $r_1a_1 + ... + r_ka_k$ ($r_i \in I$ and $a_i \in A$) is a submodule of A. Similarly, if B is a right R–module then BI is a submodule of B.

Now let R be an arbitrary ring and let B be a submodule of a left R–module A. Since B is a subgroup of the abelian group A, the factor group A/B is also defined. How does the scalar operation of R on A provide a scalar operation of R on A/B? Let a and a' be elements of A belonging to the same coset modulo B. That is to say, the difference between a and a' belongs to B. Then for an element r of R the element $r(a - a') = ra - ra'$ also belongs to B and so ra and ra' belong to the same coset modulo B. Thus ra + B = ra' + B and so there is a well–defined operation of R on A/B, given by $r \cdot (a + B) = ra + B$. We will demonstrate that this indeed defines a left R–module structure for A/B by verifying the four conditions listed in the definition of a left R–module. Less experienced readers should read these verifications carefully, giving a precise reason at each equal sign.

(1) $r[(a + B) + (a' + B)] = r[(a + a') + B]$
$$= r(a + a') + B$$
$$= (ra + ra') + B$$
$$= (ra + B) + (ra' + B)$$
$$= r(a + B) + r(a' + B).$$

(2) $(r + r')(a + B) = (r + r')a + B$
$$= (ra + r'a) + B$$
$$= (ra + B) + (r'a + B)$$
$$= r(a + B) + r'(a + B).$$

(3) $(rr')(a + B) = (rr')a + B$
$$= r(r'a) + B$$
$$= r[(r'a) + B]$$

$$= r[r'(a + B)].$$
$$(4) \quad 1(a + B) = 1a + B$$
$$= a + B.$$

The module A/B is called the *factor module* of A by B. If R is a field, then factor modules of left R–modules are just the familiar quotient spaces. If R is the ring $\mathbb{Z}$ of integers then factor modules of left $\mathbb{Z}$–modules are indistinguishable from factor groups. The role played by factor modules within the general theory of modules will appear in §9.

EXERCISES

1. Regard the ring R as a module over itself.
(a) Which subsets of R are submodules?
(b) For the special case in which R is a field, describe the submodules and factor modules of R.
(c) For the special case in which R is the ring $\mathbb{Z}$ of integers, describe the submodules and factor modules of R.

2. Let B and C be submodules of a left R–module A.
(a) Verify that $B \cap C$ is a submodule of A.
(b) Verify that the subset $B + C = \{b + c \mid b \in B$ and $c \in C\}$ is a submodule of A.

3. Give an example of a left R–module A having submodules B and C such that $B \cup C$ is not a submodule of A.

4. If $\{B_i \mid i \in \Omega\}$ is a set of submodules of a left R–module A, show that $\cap\{B_i \mid i \in \Omega\}$ is a submodule of A.

5. Let $B, C,$ and D be submodules of a left R–module A.
(a) Verify that $B \cap (C + D) = C + (B \cap D)$ whenever $B \supset C$.
(b) Show, by example, that it is not necessarily true that $B \cap (C + D)$ equals $(B \cap C) + (B \cap D)$.

§3   DIRECT PRODUCTS AND COPRODUCTS

Let R be a ring and let $\{M_j \mid j \in \Omega\}$ be a family of left R—modules indexed by an arbitrary nonempty index set Ω. From this indexed family of modules we may form a simple left R—module, $M = \Pi\{M_j \mid j \in \Omega\}$, called the *direct product* of the modules M_j. The elements of M are those functions f from Ω to the disjoint union of the M_j satisfying the condition that $f(j) \in M_j$ for all j in Ω. The sum of elements f and g of M is defined componentwise: $(f+g)(j) = f(j) + g(j)$ for all j in Ω. The scalar operation of an element r of R is defined similarly: if f is an element of M then $(rf)(j) = r[f(j)]$ for all j in Ω.

> **OBSERVATION** 1. These definitions indeed provide an R—module structure for $\Pi\{M_j \mid j \in \Omega\}$.

Whenever no ambiguity seems likely, we will denote this module in the abbreviated form ΠM_j. If Ω is empty, we define ΠM_j to be (0).

The subset N of the left R—module $\Pi\{M_j \mid j \in \Omega\}$ consisting of all those functions f for which $f(j) = 0$ for all but finitely—many elements j of Ω is called the *coproduct* (or often *external direct sum*) of $\{M_j \mid j \in \Omega\}$ and is denoted by $\amalg\{M_j \mid j \in \Omega\}$ or, when no ambiguity seems likely, simply by $\amalg M_j$.

> **OBSERVATION** 2. $\amalg M_j$ is a submodule of ΠM_j.

Note that ΠM_j and $\amalg M_j$ are equal when the set Ω is finite. If $\Omega = \{1,...,n\}$ for some positive integer n, then we often write $M_1 \amalg ... \amalg M_n$ instead of $\amalg M_j$ or ΠM_j. Direct products and coproducts of modules will play an important part in the constructions presented throughout this book.

Let us now clarify the sense in which direct products over an arbitrary index set, as defined above, may be regarded as a generalization of the elementary notion of the cartesian product. Consider left R–modules M_1 and M_2 and let $M_1 \times M_2$ be their cartesian product, i.e. the set of all ordered pairs (m_1, m_2), where m_1 is an element of M_1 and m_2 is an element of M_2. Set $\Omega = \{1, 2\}$. To fully describe an element f of ΠM_j it suffices to list the values that f assumes at 1 and at 2, keeping in mind which value is associated with 1 and which with 2. An obvious way to do this is to list the *ordered* pair $(f(1), f(2))$. This procedure of associating the ordered pair $(f(1), f(2))$ with the function f provides a one–to–one function from ΠM_j into $M_1 \times M_2$. In fact, we obtain a one–to–one correspondence between the two sets in this manner for if (m_1, m_2) is an arbitrary element of the cartesian product then it is associated with the element g of ΠM_j for which $g(1) = m_1$ and $g(2) = m_2$. Notice that if f and g are elements of ΠM_j and if r is an element of R then f + g corresponds to the ordered pair $((f+g)(1), (f+g)(2)) = (f(1)+g(1), f(2)+g(2))$ and rf corresponds to the ordered pair $((rf)(1), (rf)(2)) = (rf(1), rf(2))$. Therefore the operations on ΠM_j correspond to the (familiar) coordinatewise operations on ordered pairs. Because of all this, the sets $M_1 \amalg M_2$ and $M_1 \times M_2$ are often identified and elements of $M_1 \amalg M_2$ are written as ordered pairs, with componentwise operations.

If Λ is a subset of a set Ω and if $\{M_j \mid j \in \Omega\}$ is a family of left R–modules then any function f in $\Pi\{M_j \mid j \in \Lambda\}$ can be extended in a natural way to a function f' in $\Pi\{M_j \mid j \in \Omega\}$ defined by $f'(j) = f(j)$ if j is in Λ and $f'(j) = 0$ otherwise. This allows us to identify $\Pi\{M_j \mid j \in \Lambda\}$ with a submodule of $\Pi\{M_j \mid j \in \Omega\}$ and, similarly, to identify $\amalg\{M_j \mid j \in \Lambda\}$ with a submodule of $\amalg\{M_j \mid j \in \Omega\}$.

EXERCISES

1. Verify Observation 1.

2. Verify Observation 2.

§4 HOMOMORPHISMS OF MODULES

Let A and B be left R–modules. A function α from
A to B which assigns to each element a of A an element
aα of B is called an R-*homomorphism* if and only if (a
+ a')α = aα + a'α and (ra)α = r(aα) for all a and a' in
A and r in R. (The reason for writing α to the right of its
operand will become clear in the next section.) When it is clear
which ring is involved, we will often use the term *map* instead
of the more cumbersome "R–homomorphism". Similarly, we can
define the notion of an R–homomorphism of right R–modules.
Such maps are written as acting on the left. The rule is very
simple: module homomorphisms always act on the side opposite
scalar multiplication.

If R is a field then the R–homomorphisms are just the
familiar R–linear transformations studied in linear algebra. If
A and B are abelian groups then a function from A to B
is a group homomorphism precisely when it is a
$\mathbb{Z}$–homomorphism from A to B when these are regarded as
left $\mathbb{Z}$–modules. Without this fact, we could not fully identify,
as we do, the theory of abelian groups and the theory of left
$\mathbb{Z}$–modules. The word "homomorphism" is derived from the
Greek words meaning "same structure" and is used in many
situations in algebra to denote a function the action of which
preserves the algebraic structure we consider important.
Therefore, be careful: there is an important difference between
homomorphisms of left R–modules, as we have just defined
them, and homomorphisms of rings (see Exercise 1). In this

book, we will use module homomorphisms constantly, but make only occasional use of ring homomorphisms.

If α is an R−homomorphism from a left R−module A to a left R−module B and if β is an R−homomorphism from B to a left R−module C then, as is easily verified, the composition of these two maps is an R−homomorphism from A to C. In keeping with our convention of writing homomorphisms of left modules as acting on the right, we will denote this composition by $\alpha\beta$. On the other hand, if A, B, and C are right R−modules then the composition of α and β is a homomorphism written as acting on the left and so is denoted by $\beta\alpha$. All this may seem a bit confusing at first, but, as we shall see in the next section, it is very natural.

If α is an R−homomorphism from a left R−module A to a left R−module B then the subset $\{a \in A \mid a\alpha = 0\}$ of A is called the *kernel* of α and is denoted by $\ker(\alpha)$.

OBSERVATION 1. If $\alpha{:}A \to B$ is an R−homomorphism between left R−modules then $\ker(\alpha)$ is a submodule of A.

If $\ker(\alpha) = (0)$, then α is said to be an R−*monomorphism*. An R−homomorphism has the property that distinct elements of its domain have distinct images (that is to say, it is one−to−one) if and only if it is a monomorphism. The subset $\{a\alpha \mid a \in A\}$ of B is called the *image* of α and may be denoted by $\text{im}(\alpha)$ or by $A\alpha$.

OBSERVATION 2. If $\alpha{:}A \to B$ is an R−homomorphism of left R−modules then $\text{im}(\alpha)$ is a submodule of B.

If $\text{im}(\alpha) = B$ then α is said to be an R−*epimorphism*. An R−homomorphism is surjective (that is to say, onto) if and only if it is an epimorphism. If α is both a monomorphism and an epimorphism then it is said to be an R−*isomorphism*.

An important special case occurs when we consider maps from a left R−module to itself. An R−homomorphism from a left R-module A to itself is called an R-*endomorphism* of A. An R–isomorphism from A to itself is called an R-*automorphism* of A. For any left R-module A there always exists at least one R–automorphism of A, namely the *identity map* on A which sends every element of A to itself. If A is not (0) then there exists at least one R–endomorphism of A which is not an R–automorphism, namely the *zero map* on A which sends every element of A to the element 0.

If B is a submodule of a left R–module A then B defines two special R–homomorphisms. The first of these is the *inclusion map* $\iota_{B,A}:B \to A$ which is defined by $b\iota_{B,A} = b$ for all b in B. Notice that the identity map on A is merely $\iota_{A,A}$. The second is the *natural map* $\nu_{B,A}:A \to A/B$ defined by $a\nu_{B,A} = a + B$ for all a in A.

Our next objective is to show that there is a way of regarding all module homomorphisms as special types of submodules, a fact that will prove useful in Chapters 6 and 7. If α is an R–homomorphism from a left R–module A to a left R–module B then the *graph* of α is the subset gr(α) = $\{(a,a\alpha) \mid a \in A\}$ of the cartesian product $A \times B$. In §3 we already noted that $A \times B$ can be identified with the left R–module $A \amalg B$ and so can be considered as a left R–module. Then gr(α) becomes a submodule of $A \times B$ since $(a,a\alpha) + (a',a'\alpha) = (a+a',a\alpha+a'\alpha) = (a+a',(a+a')\alpha)$ and $r(a,a\alpha) = (ra,ra\alpha) = (ra,(ra)\alpha)$ are in gr(α) for all a and a' in A and all r in R. Distinct maps from A to B have distinct submodules of $A \times B$ as graphs. Which submodules of $A \times B$ are graphs of maps from A to B? To answer this question we need to observe one more property of gr(α): Let $\pi:A \times B \to A$ be the function which assigns to each ordered pair (a,b) its first component, a. Since α is a function with domain A, the restriction π' of π to gr(α) yields a one-to-one correspondence between gr(α) and A. This last property characterizes the submodules of $A \times B$ which are graphs of maps: a submodule G of $A \times B$ is the graph of an R–homomorphism α from A to B if and only if the restriction of π to G is a one–to–one correspondence between G and A. Given such a submodule G, it is easy to construct the required map α––for each a in A, there is exactly one pair in G of the form (a,b), so set $a\alpha = b$. That this α is in fact a map follows from the fact that S is a submodule of A $\times$ B.

If A and B are left R–modules then the set of all R–homomorphisms from A to B will be denoted by $Hom_R(A,B)$. This set is nonempty since it always contains the zero map which sends every element of A to the additive identity element of B. If α and β are R–homomorphisms from A to B then the function $\alpha + \beta$ from A to B defined by $a(\alpha + \beta) = a\alpha + a\beta$ for all a in A is also an R–homomorphism from A to B since $(a + a')(\alpha + \beta) = (a + a')\alpha + (a + a'')\beta = a\alpha + a'\alpha + a\beta + a'\beta = a(\alpha + \beta) + a'(\alpha + \beta)$ and $(ra)(\alpha + \beta) = (ra)\alpha + (ra)\beta = r(a\alpha) + r(a\beta) = r[a(\alpha + \beta)]$ for all a and a' in A and r in R.

OBSERVATION 3. If A and B are left R–modules then $Hom_R(A,B)$ is an additive abelian group.

Now let us consider the special case in which B equals R itself. If A is any left R–module then $A^* = Hom_R(A,R)$ is, as we have seen, an additive abelian group. For any element α of this group and for any element r of R, let αr be the function from A to R which sends an element a to $(a\alpha)r$. This is in fact a homomorphism of left R–modules for if a and a' are elements of A and if r' is an element of R then $[(a+a')\alpha]r = (a\alpha + a'\alpha)r = (a\alpha)r + (a'\alpha)r$ and $[(r'a)\alpha]r = [r'(a\alpha)]r = r'[(a\alpha)r]$. Thus αr belongs to A^*. It fact, it is easily seen, given this definition and Observation 3, that

OBSERVATION 4. If A is a left R–module then $A^* = Hom_R(A,R)$ is a right R–module.

Similarly, if A is a right R–module then $A^* = \text{Hom}_R(A,R)$ is a left R–module.

EXERCISES

1. Regard the field $\mathbb{C}$ of complex numbers as a module over itself.
(a) Verify that the function $\varphi:\mathbb{C} \to \mathbb{C}$ defined by $\varphi(a+bi) = a-bi$ is a ring homomorphism but not a $\mathbb{C}$–module homomorphism.
(b) Verify that the function $\psi:\mathbb{C} \to \mathbb{C}$ defined by $\psi(a+bi) = 2a+2bi$ is a $\mathbb{C}$–module homomorphism but not a ring homomorphism.

2. If $\alpha:A \to B$ is a map of left R–modules, verify that ker(α) is indeed a submodule of A and that im(α) is indeed a submodule of B.

3. Let A and B be left R–modules and let G be a submodule of the left R–module $A \times B$ having the property that for each a in A there is a unique element b of B such that (a,b) is in G. Verify that the function α from A to B defined by the condition that $a\alpha = b$ provided that $(a,b) \in G$ is indeed an R–homomorphism.

4. If A and B are left R–modules, verify that the set $\text{Hom}_R(A,B)$ is an additive abelian group.

5. Let $\alpha_1:A \to B$, $\alpha_2:A \to B$, and $\beta:B \to C$ be maps of left R–modules for which β is a monomorphism and $\alpha_1\beta = \alpha_2\beta$. Show that $\alpha_1 = \alpha_2$.

6. Let $\alpha:A \to B$, $\beta_1:B \to C$, and $\beta_2:B \to C$ be maps of left R–modules for which α is an epimorphism and $\alpha\beta_1 = \alpha\beta_2$. Show that $\beta_1 = \beta_2$.

7. Let B' be a submodule of a left R–module B and let $\lambda : B' \to B$ be the inclusion map. For an arbitrary left R–module A,

(a) Show that the function $\varphi_{B'}$ from $\operatorname{Hom}_R(A,B')$ to $\operatorname{Hom}_R(A,B)$ which sends each map α to the composite map $\alpha\lambda$ is a homomorphism of abelian groups.

(b) Show that there is no proper submodule B' of B such that $\varphi_{B'}$ is an isomorphism if and only if B is isomorphic to a factor module of a coproduct of copies of A.

8. Let B' be a submodule of a left R–module B and let $\pi : B \to B/B'$ be the natural map. For an arbitrary left R–module A,

(a) Show that the function $\psi_{B'}$ from $\operatorname{Hom}_R(B/B',A)$ to $\operatorname{Hom}_R(B,A)$ which sends each map α to the composite map $\pi\alpha$ is a homomorphism of abelian groups.

(b) Show that there is no nonzero submodule B' of B such that $\psi_{B'}$ is an isomorphism if and only if B is isomorphic to a submodule of a direct product of copies of A.

9. If R is a ring, let $Z(R) = \{r' \in R \mid r'r = rr'$ for all r in R$\}$. This set is called the *center* of R.

(a) Show that $Z(R)$ is a commutative subring of R.

(b) If A and B are arbitrary left R–modules, show that $\operatorname{Hom}_R(A,B)$ can be made into a left $Z(R)$–module in a natural way.

10. Show that the left $\mathbb{Z}$–modules $\mathbb{Q}$ and $\operatorname{Hom}_{\mathbb{Z}}(\mathbb{Q},\mathbb{Q})$ are isomorphic.

§5 BIMODULES

In §1 we defined left and right modules over a ring R. Actually, as we shall now see, every left R–module is a right

S—module for some ring S and every right R—module is similarly a left S—module. We will want these module structures to coexist nicely.

DEFINITION. If R and S are rings then an (R,S)-*bimodule* is an additive abelian group A satisfying the following conditions:
(1) A is a left R—module,
(2) A is a right S—module, and
(3) r(as) = (ra)s
for all elements r of R, s of S, and a of A.

In the example mentioned above, if A is a vector space of dimension n over a field F and if S is the full ring of n$\times$n matrices over F then A is an (F,S)—bimodule. Similarly, any ring R is an (R,R)—bimodule.

How do these examples generalize? By Observation 3 of §3, we see that if A is a left R—module then $Hom_R(A,A)$, the set of all R—endomorphisms of A, is an additive abelian group. There is also another operation definable on this set, namely composition of functions: if α and β are R—endomorphisms of A then $\alpha\beta$ is the R—endomorphism of A defined by $a(\alpha\beta) = (a\alpha)\beta$. We then make two very important observations:

OBSERVATION 1. If A is a left R—module then $Hom_R(A,A)$ is a ring under the operations of homomorphism addition and composition. The multiplicative identity of this ring is precisely the identity map $\iota_{A,A}$.

OBSERVATION 2. If A is a left R—module and if S = $Hom_R(A,A)$ then A is an (R,S)—bimodule.

Similarly, if A is a right R—module and if S = $Hom_R(A,A)$ then A is an (S,R)—bimodule.

OBSERVATION 3. If A is an (R,S)—bimodule then

$Hom_R(R,A)$ is a right S−module isomorphic to A.

VERIFICATION: We will only sketch the argument here, leaving the details as an exercise. If $\alpha \in Hom_R(R,A)$ and if $s \in S$, define αs to be the function from R to A which sends an element r to $(r\alpha)s$. This is a map from R to A and, with this definition of scalar multiplication by elements of S, $Hom_R(R,A)$ is a right S−module. Now consider the *evaluation function* ε from $Hom_R(R,A)$ to A defined as follows: if α is a map from R to A then $\varepsilon(\alpha) = 1\alpha$. Then ε is a map of right S−modules. If $a \in A$ and if α_a is the map from R to A defined by $r\alpha_a = ra$ for all r in R, then $\varepsilon(\alpha_a) = a$ and so ε is an epimorphism. If $\varepsilon(\alpha) = 0$ then $(r\alpha) = (r1\alpha) = r(1\alpha) = r0 = 0$ for all r in R and so α is the zero map. Thus ε is a monomorphism and so it is an isomorphism. □

A similar argument shows that if A is an (R,S)−bimodule then $Hom_S(S,A)$ is a left R−module isomorphic to A.

EXERCISES

1. Complete the details of the verification of Observation 3.

2. Let R and S be rings, let A be a left R−module, and let B be a (R,S)−bimodule. If $\alpha \in Hom_R(A,B)$ and if s is an element of S, define the map αs from A to B by $a(\alpha s) = (a\alpha)s$ for all a in A. Show that αs belongs to $Hom_R(A,B)$ and that, in this way, we have turned $Hom_R(A,B)$ into a right S−module.

3. Let R and S be rings, let A be an (R,S)−bimodule, and let B be a left R−module. If $\alpha \in Hom_R(A,B)$ and if s is an element of S, define the map $s\alpha$ from A to B by $a(s\alpha) = (as)\alpha$ for all a in A. Show that $s\alpha$ belongs to $Hom_R(A,B)$ and that, in this way, we have turned $Hom_R(A,B)$ into a left S−module.

4. Let R and S be rings and let A be an (R,S)-bimodule. Let T be the set of all matrices of the form $\begin{bmatrix} r & a \\ 0 & s \end{bmatrix}$ with $r \in$ R, $s \in$ S, and $a \in$ A. Define addition and multiplication in T as follows: if $x = \begin{bmatrix} r & a \\ 0 & s \end{bmatrix}$ and $y = \begin{bmatrix} r' & a' \\ 0 & s' \end{bmatrix}$ are elements of T then $x + y = \begin{bmatrix} r+r' & a+a' \\ 0 & s+s' \end{bmatrix}$ and $xy = \begin{bmatrix} rr' & ra'+as' \\ 0 & ss' \end{bmatrix}$.

(a) Show that T with the operations defined is a ring.

(b) Show that the set H of all elements of the ring T of the form $\begin{bmatrix} 0 & a \\ 0 & 0 \end{bmatrix}$ (which we can identify with A) is an ideal of R.

(c) Show that $H^2 = (0)$.

(d) Show that the ring $R \times S$ is isomorphic to a subring of T.

(e) Show that every left ideal of T has the form $\{\begin{bmatrix} r & a \\ 0 & s \end{bmatrix} \mid (r,a) \in X, s \in Y\}$, where X is a submodule of the left R−module R $\amalg$ A (considered as a set of ordered pairs) and Y is a left ideal of S satisfying $AY \subseteq X$.

§6   PROJECTIONS AND INJECTIONS

Let $\{M_j \mid j \in \Omega\}$ be a family of modules over a ring R, and let ΠM_j be the direct product of these modules. With each h in Ω we associate a function $\pi_h : \Pi M_j \to M_h$ defined by evaluating each function f in ΠM_j at h. Thus $f\pi_h = f(h)$ for each $f \in \Pi M_j$. The function π_h is in fact an R−homomorphism since $(f + f')\pi_h = (f + f')(h) = f(h) + f'(h) = f\pi_h + f'\pi_h$ and $(rf)\pi_h = (rf)(h) = r[f(h)] = r(f\pi_h)$ for all f and f' in ΠM_j and all r in R. The map π_h is called the *projection map* onto the hth coordinate of ΠM_j.

Projection maps are used in the following common and important construction: Let A be a left R−module and let $\{M_j \mid j \in \Omega\}$ be a family of left R−modules. For each j in Ω, let $\alpha_j : A \to M_j$ be an R−homomorphism. Then from the family $\{\alpha_j\}$ of maps we can construct a single map $\alpha : A \to$

ΠM_j by associating with each element a in A the function f_a in ΠM_j defined by $f_a(j) = a\alpha_j$ for all j in Ω. To show that α is indeed an R–homomorphism, we must show that for all a and a' in A and all r in R the function $(a+a')\alpha$ is the same as $a\alpha + a'\alpha$ and the function $(ra)\alpha$ is the same as $r(a\alpha)$. To show that two functions are identical, we must show that they have the same effect when applied to an element in their common domain. Thus, indeed, if j is an element of Ω then $((a+a')\alpha)(j) = f_{a+a'}(j) = (a+a')\alpha_j = a\alpha_j + a'\alpha_j = f_a(j) + f_{a'}(j) = (a\alpha)(j) + (a'\alpha)(j) = (a\alpha + a'\alpha)(j)$. Likewise, if j is an element of Ω then $((ra)\alpha)(j) = f_{ra}(j) = (ra)\alpha_j = r(a\alpha_j) = rf_a(j) = (r(a\alpha))(j)$. Note that if j is an element of Ω and if $a \in A$ then $a\alpha\pi_j = (f_a)\pi_j = f_a(j) = a\alpha_j$ and so $\alpha_j = \alpha\pi_j$. This property completely characterizes the map α, for if there exists a map β from A to ΠM_j satisfying the condition that $\beta\pi_j = \alpha_j$ for all j in Ω then for any a in A we would have $(a\beta)(j) = (a\beta\pi_j) = a\alpha_j = (a\alpha)(j)$ for all j in Ω, proving that $a\beta = a\alpha$ and thus $\beta = \alpha$. We summarize this construction in the following observation:

OBSERVATION 1. If A is a left R–module and if $\{M_j \mid j \in \Omega\}$ is a family of left R–modules such that for each j in Ω there exists an R–homomorphism $\alpha_j{:}A \to M_j$ then there exists an R–homomorphism $\alpha{:}A \to \Pi M_j$ satisfying the condition that $\alpha_j = \alpha\pi_j$ for each j in Ω. Moreover, the map having this property is unique.

There is a second family of maps associated with the direct product construction. If $\{M_j \mid j \in \Omega\}$ is a family of left R–modules then for each h in Ω let $\lambda_h{:}M_h \to \Pi M_j$ be the function defined as follows: if $m \in M_h$ then $m\lambda_h(j)$ equals m if $j = h$ and equals 0 otherwise. To show that λ_h is an R–homomorphism we must show that for all m and m' in M_h and for all r in R, the function $(m + m')\lambda_h$ is the same as $m\lambda_h + m'\lambda_h$ and the function $(rm)\lambda_h$ is the same as the function $r(m\lambda_h)$. Since all four of these functions take the value 0 except at h, the following computations are enough to prove our claim: $((m + m')\lambda_h)(h) = m + m' = (m\lambda_h)(h) + (m'\lambda_h)(h) = (m\lambda_h + m'\lambda_h)(h)$ and $((rm)\lambda_h)(h) = rm =(r(m\lambda_h))(h)$. The map λ_h is called the *injection map* into the hth coordinate of ΠM_j.

Note that for each h in Ω the injection map into the hth coordinate of ΠM_j is an R–monomorphism, and so M_h is isomorphic to $im(\lambda_h)$. This allows us, sometimes, to identify M_h with its image under λ_h and so consider M_h as a submodule of ΠM_j.

Thus with each direct product $\Pi\{M_j \mid j \in \Omega\}$ of left R–modules we have associated two fundamental families of maps: projections π_j and injections λ_j. Each injection λ_j can be composed with each projection π_k. Composites of the form $\lambda_j\pi_j$ yield identity maps since, for each x in M_j, $x(\lambda_j\pi_j) = (x\lambda_j)\pi_j = (x\lambda_j)(j) = x$. Composites of the form $\lambda_j\pi_k$ for $j \neq k$ yield zero maps since, for every x in M_j, $x\lambda_j\pi_k = (x\lambda_j)\pi_k = (x\lambda_j)(k) = 0$. The fact that $\lambda_j\pi_j$ is the identity map on M_j encompasses two obvious but important facts: (1) the projections π_j are epimorphisms, and (2) the injections λ_j are monomorphisms. In accord with the informal usage as described above, $im(\lambda_j) = M_j$ and $ker(\pi_j) = \Pi\{M_i \mid i \in \Omega \setminus \{j\}\}$.

When $\Omega = \{1,2\}$, we may regard the elements of $\Pi M_i = M_1 \amalg M_2$ as ordered pairs, as explained in §3. Our constructions and computations are especially transparent in this case. The projections and injections are exhaustively described by $(m_1,m_2)\pi_1 = m_1$, $(m_1,m_2)\pi_2 = m_2$, $m_1\lambda_1 = (m_1,0)$, and $m_2\lambda_2 = (0,m_2)$, where m_1 and m_2 are arbitrary elements of M_1 and M_2 respectively. For each pair of maps $\alpha_1:A \to M_1$ and $\alpha_2:A \to M_2$ there exists a map $\alpha:A \to M_1 \amalg M_2$ satisfying $\alpha_1 = \alpha\pi_1$ and $\alpha_2 = \alpha\pi_2$. The map α admits a simple description: $a\alpha = (a\alpha_1, a\alpha_2)$ for each a in A. The informality of identifying M_i with $im(\lambda_i)$ amounts to blurring the distinction between M_1 and $\{(m_1,0) \mid m_1 \in M_1\}$ and between M_2 and $\{(0,m_2) \mid m_2 \in M_2\}$. The computation of such composites as $\lambda_1\pi_1$ and $\lambda_1\pi_2$ can be illustrated by $m_1\lambda_1\pi_1 = (m_1,0)\pi_1 = m_1$ and $m_1\lambda_1\pi_2 = (m_1,0)\pi_2 = 0$.

Let $\{M_i \mid i \in \Omega\}$ be a family of left R–modules, let $\Pi\{M_i \mid i \in \Omega\}$ be the direct product of this family, and let $\{\pi_i \mid i \in \Omega\}$ and $\{\lambda_i \mid i \in \Omega\}$ be associated families of projections and injections as described above. Inside ΠM_i we have the submodule $\amalg\{M_i \mid i \in \Omega\}$. The restriction of each projection π_h to $\amalg M_i$ results in a map from $\amalg M_i$ to M_h which, to avoid cumbersome notation, we will also denote by π_h. No confusion will arise as a result of this shorthand because in dealing with such maps we will always make the domain clear. Since $im(\lambda_h) \subseteq \amalg M_i$ for each h in Ω, λ_h can be considered as a map from M_h to $\amalg M_j$.

Let A be a fixed left R–module and let $\{\alpha_i:M_i \to A \mid i \in \Omega\}$ be a family of R–homomorphisms. Then from this family we can define a single function α from $\amalg M_i$ to A by associating with each f in $\amalg M_i$ the sum of the (finite!) set $\{f(j)\alpha_j) \mid f(j) \neq 0\}$ of elements of A. Straightforward computations, left as an exercise, show us that α is in fact an R–homomorphism, which will be denoted by $\Sigma\alpha_i$. If $\lambda_h:M_h \to \amalg M_i$ is the injection of M_h into $\amalg M_i$ then $\lambda_h(\Sigma\alpha_i) = \alpha_h$ for all h in Ω.

In particular, we can consider a family $\{A_i \mid i \in \Omega\}$ of submodules of a left R–module A. For each i in Ω, let $\alpha_i:A_i \to A$ be the inclusion map. By the previous discussion we see that these maps define an R–homomorphism $\alpha:\amalg A_i \to A$ the image of which is a submodule of A which is denoted by $\Sigma\{A_i \mid i \in \Omega\}$.

Now let us restrict our consideration to the special case of the coproduct of a pair A, B of left R–modules. For simplicity, we will denote the projections of A $\amalg$ B onto A and B respectively by π_A and π_B and the injections of A and B respectively into A $\amalg$ B by λ_A and λ_B. If C is a left R–module and if $\alpha_A:A \to C$ and $\alpha_B:B \to C$ are maps, then there exists a unique map $\alpha: A \amalg B \to C$ satisfying the conditons that $\lambda_A \alpha = \alpha_A$ and $\lambda_B \alpha = \alpha_B$, namely the map α which assigns to the pair (a,b) the value $a\alpha_A + b\alpha_B$. The remarks in §4 concerning the graphs of maps may be reformulated in our present terminology: a submodule G of A $\amalg$ B is the graph of a map $\alpha: A \to B$ if and only if the restriction of π_A to G is an isomorphism. A submodule G' of A $\amalg$ B for which the restriction of π_A to G' is one-to-one may be regarded as the graph of a map from $G'\pi_A$ to B.

Let B be a left R-module and let $\{A_i \mid i \in \Omega\}$ be a family of left R-modules. Then we have a function θ from $\text{Hom}_R(\amalg A_i, B)$ to $\Pi \text{Hom}_R(A_i, B)$ defined by $(\alpha\theta)(h) = \lambda_h \alpha$ for each α in $\text{Hom}_R(\amalg A_i, B)$ and each h in Ω. It is left as an exercise to show that this is in fact an isomorphism of abelian groups, and so to verify the following observation.

OBSERVATION 2. Let B be a left R-module and let $\{A_i \mid i \in \Omega\}$ be a family of left R-modules. Then $\text{Hom}_R(\amalg A_i, B) \cong \Pi \text{Hom}_R(A_i, B)$. Moreover, if B is an (R,S)-bimodule then this is an isomorphism of right S-modules.

Similarly, let B be a left R-module and let $\{A_i \mid i \in \Omega\}$ be a family of left R-modules. Then we have a function ψ from $\text{Hom}_R(B, \Pi A_i)$ to $\Pi \text{Hom}_R(B, A_i)$ defined by $\psi(\alpha)(h) = \alpha\pi_h$ for each α in $\text{Hom}_R(B, \Pi A_i)$ and each h in Ω. Again, it is left as an exercise to show that this is in fact an isomorphism of abelian groups and so to verify the following observation.

OBSERVATION 3. Let B be a left R-module and let $\{A_i \mid i \in \Omega\}$ be a family of left R-modules. Then $\text{Hom}_R(B, \Pi A_i) \cong \Pi \text{Hom}_R(B, A_i)$. Moreover, if B is an (R,S)-bimodule then this is an isomorphism of left S-modules.

EXERCISES

1. Let $\{B_i \mid i \in \Omega\}$ be a family of left R–modules and let $B = \Pi B_i$. Let α and β be R–homomorphisms from a left R–module A to B for which $\alpha \pi_j = \beta \pi_j$ for every projection $\pi_j : B \to B_j$ $(j \in \Omega)$. Show that $\alpha = \beta$.

2. Let A be a fixed left R–module and let $\{\alpha_i : M_i \to A \mid i \in \Omega\}$ be a family of R–homomorphisms. Verify that the function $\Sigma \alpha_i$ from $\amalg M_i$ to A is an R–homomorphism.

3. Verify Observation 2.

4. Verify Observation 3.

5. Let B be a left R–module and let $\{A_i \mid i \in \Omega\}$ be a family of left R–modules. Define a function φ from the abelian group $\amalg\{Hom_R(B,A_i) \mid i \in \Omega\}$ to the abelian group $Hom_R(B, \amalg A_i)$ as follows: if $f \in \amalg Hom_R(B,A_i)$ then $\varphi(f) = \Sigma f(i)$.
(a) Show that φ is a homomorphism of abelian groups.
(b) Provide an example to show that it is not necessarily an isomorphism.
(c) If B is an (R,S)–bimodule, show that φ is a homomorphism of left S–modules.

6. Let B, $\{A_i \mid i \in \Omega\}$, and φ be as in Exercise 2. Show that a necessary and sufficient condition for φ to be an isomorphism is that the union of any ascending chain $B_1 \subset B_2 \subset ...$ of proper submodules of B is again a proper submodule of B.

7. Let A be a left R–module, let n be a positive integer, and let B be a coproduct of n copies of A. Moreover, let $S = Hom_R(A,A)$ be the ring of endomorphisms of A.
(a) Show that the ring $Hom_R(B,B)$ is isomorphic to the full

ring of $n \times n$ matrices over S.
(b) Note that B can also be considered as a left module over the full ring T of $n \times n$ matrices over R. Show that the ring $\text{Hom}_T(B,B)$ is isomorphic to S.

8. Let M_1 and M_2 be left R–modules and let $M = M_1 \amalg M_2$; for each $i = 1,2$, let $\lambda_i : M_i \to M$ be the injection map. Suppose that we are given a left R–module N and R–homomorphisms $\alpha_i : N \to M_i$ for $i = 1,2$.
(a) Show that $M' = \{(n\alpha_1, -n\alpha_2) \mid n \in N\}$ is a submodule of M.
(b) If $\nu : M \to M/M'$ is the natural map and if $\beta_i = \lambda_i \nu$ for $i = 1,2$, show that $\alpha_1 \beta_1 = \alpha_2 \beta_2$.
(c) If A is a left R–module and if $\gamma_i : M_i \to A$ $(i = 1,2)$ are R–homomorphisms satisfying $\alpha_1 \gamma_1 = \alpha_2 \gamma_2$, show that there exists a unique R–homomorphism $\theta : M/M' \to A$ satisfying $\beta_i \theta = \gamma_i$ for $i = 1,2$.

9. Let M_1 and M_2 be left R–modules and let $M = M_1 \amalg M_2$; for each $i = 1,2$, let $\pi_i : M \to M_i$ be the projection map. Suppose that we are given a left R–module N and R–homomorphisms $\alpha_i : M_i \to N$ for $i = 1,2$.
(a) Show that $M' = \{(m_1, m_2) \mid m_1 \alpha_1 = m_2 \alpha_2\}$ is a submodule of M.
(b) If $\iota : M' \to M$ is the inclusion map and if $\beta_i = \iota \pi_i$ for $i = 1,2$, show that $\beta_1 \alpha_1 = \beta_2 \alpha_2$.
(c) If A is a left R–module and if $\gamma_i : A \to M_i$ $(i = 1,2)$ are R–homomorphisms satisfying $\gamma_1 \alpha_1 = \gamma_2 \alpha_2$, show that there exists a unique R–homomorphism $\theta : A \to M'$ satisfying $\gamma_i = \theta \beta_i$ for $i = 1,2$.

§7 INTERNAL DIRECT SUMS

A constant theme of our work will be the attempt to represent an R–module A of some specified type as a coproduct of modules of some structurally more transparent type. However, the appropriate question for us to ask can *not* be "Is there a family of R–modules $\{B_i \mid i \in \Omega\}$ for which A equals $\amalg B_i$?" because the elements of the set on the right are of a specific nature: they are functions from Ω into $\cup B_i$. Thus (except when the elements of A happen to be such functions), A has no chance of being *equal* to $\amalg B_i$. Our question will have to be phrased "Is there a family of R–modules $\{B_i \mid i \in \Omega\}$ for which A is isomorphic to $\amalg B_i$?" This makes life seem a bit more difficult, since we do not know where to look for the modules B_i. We would therefore like to come up with an equivalent question which is "internal" to A, i.e. which is involved with finding a certain family of submodules of the module A.

OBSERVATION 1. A left R–module A is isomorphic to a coproduct $\amalg\{B_i \mid i \in \Omega\}$ of left R–modules if and only if there exists a family $\{A_i \mid i \in \Omega\}$ of submodules of A satisfying the following conditions:

(1) $A_i \cong B_i$ for each i in Ω.

(2) For each element a of A there is a unique indexed family $\{a_i \mid i \in \Omega\}$ of elements of A that satisfies the following conditions:

    (a) $a_i \in A_i$ for all i in Ω;

    (b) $a_i = 0$ for all but finitely–many i in Ω;

    (c) $a = \Sigma\{a_i \mid i \in \Omega\}$.

VERIFICATION: Suppose that A contains a family of submodules $\{A_i \mid i \in \Omega\}$ that satisfies conditions (1) and (2). For each h in Ω, let $\lambda_h:B_h \rightarrow \amalg B_i$ be the injection of B_h into the coproduct of the B_i and let $\alpha_h:B_h \rightarrow A$ be the map which is the composite of an isomorphism $B_h \rightarrow A_h$ (which exists by condition (1)) and the inclusion map $A_h \rightarrow A$. Define the map $\alpha:\amalg B_i \rightarrow A$ to be $\Sigma\alpha_i$.

We will show that α is an isomorphism. Let f be in the

kernel of α. Then $0 = \Sigma f(i)\alpha_i$. We will use the uniqueness
assumed in condition (2) as follows. Let $\{z_i \mid i \in \Omega\}$ be the
family of elements of A defined by $z_i = 0$ for all $i \in \Omega$.
Then each of the two families $\{f(i)\alpha_i\}$ and $\{z_i\}$ is finitely
nonzero and has sum 0. By the uniqueness criterion in
condition (2), these families must be identical and so $f(i)\alpha_i = 0$
for all i in Ω. Since each α_i is a monomorphism, this means
that $f(i) = 0$ for each i in Ω and so f is the zero element
of IIB_i. Thus α is a monomorphism. To show that it is an
epimorphism, we let a be an arbitrary element of A and
construct an element f of IIB_i as follows. By condition (2),
we have $a = \Sigma a_i$ with each a_i in the image of α_i and so
for each i in Ω there exists an element b_i of B_i such that
$a = \Sigma b_i\alpha_i$. Moreover, all but finitely—many of these b_i are
zero. Now let f be the function in IIB_i defined by $f(i) = b_i$
for all i. Then $f\alpha = \Sigma b_i\alpha_i = a$, showing that α is an
epimorphism and hence an isomorphism.

Now suppose that there is an isomorphism α from IIB_i
to A. For each i, let $A_i = B_i\alpha$. This is a submodule of
A and it is left as an exercise to show that the family $\{A_i\}$ of
submodules of A satisfies condition (2). $\square$

If $\{A_i \mid i \in \Omega\}$ is a family of submodules of a left
R—module A which satisfies condition (2) of Observation 1, then
A is said to be the *(internal) direct sum* of the A_i, and
we write $A = \oplus A_i$. It is necessary and important to
distinguish between the notion of the direct sum *of a family
of submodules of a given module* and that of the
coproduct *of an arbitrary family of modules*. These
concepts, however, are closely related; as we have seen, if a left
R—module is isomorphic to a coproduct of a family of modules
then it is a direct sum of submodules isomorphic to members of
this family. Therefore many authors use the same notation "$\oplus$"
for both direct sums and coproducts. No harm comes of this as
a rule, so long as one makes clear what is being intended at
each step.

We will often be interested in showing that a module is the
direct sum of two of its submodules. In that case, a slight

variation on the verification of Observation 1 yields the following result.

> **OBSERVATION** 2. Let B and C be submodules of a left R–module A. Then $A = B \oplus C$ if and only if $B \cap C = (0)$ and for each a in A there are elements b in B and c in C for which $a = b + c$.

EXERCISES

1. Complete the verification of Observation 1.

2. Verify Observation 2.

3. Let $A \subseteq B \subseteq C$ be left R–modules and let A be a submodule of C satisfying $C = A \oplus A'$. Show that $B = A \oplus (B \cap A')$.

§8 SPLITTING MAPS AND SUMMANDS

The projection π_A and the injection λ_A associated with a coproduct $A \amalg B$ have the following two properties: the composite map $\lambda_A \pi_A$ is the identity map on A and $A \oplus B$ is the direct sum of $\mathrm{im}(\lambda_A)$ and $\ker(\pi_A)$ ($= \mathrm{im}(\lambda_B)$). A comparison of these two properties of π_A and λ_A motivates the following observation.

> **OBSERVATION** 1. Let $\alpha{:}A \to B$ and $\beta{:}B \to A$ be homomorphisms between left R–modules A and B for which $\alpha\beta$ is the identity map on A. Then $B = \mathrm{im}(\alpha) \oplus \ker(\beta)$.

VERIFICATION: Let $x \in \text{im}(\alpha) \cap \text{ker}(\beta)$. Then $x = a\alpha$ for some a in A and also $x\beta = 0$. Consequently, $a = a\alpha\beta = x\beta = 0$ and $x = a\alpha = 0\alpha = 0$. Thus $\text{im}(\alpha) \cap \text{ker}(\beta) = (0)$. Let $b \in B$. Then $b\beta\alpha \in \text{im}(\alpha)$ and $b = b\beta\alpha + (b - b\beta\alpha)$. To complete the verification, we need only show that $b - b\beta\alpha$ is in $\text{ker}(\beta)$, which is so since $(b - b\beta\alpha)\beta = b\beta - b\beta\alpha\beta = b\beta - b\beta = 0$. $\square$

When maps $\alpha{:}A \to B$ and $\beta{:}B \to A$ have the property that $\alpha\beta$ is the identity map, we say that each of the maps is a *splitting map* for the other. As Observation 1 shows, splitting maps are valuable tools for splitting modules into internal direct sums. Notice that, when $\alpha\beta$ is the identity map, it follows that α is monic and β is epic. The maps π_A and λ_A mentioned above are splitting maps of each other.

A submodule B of a left R–module A is a *direct summand* of A if there is a submodule C of A for which $A = B \oplus C$. When such a C exists, it is said to be a *complementary summand* of B. Any left R–module A can be written as $A \oplus (0)$, and consequently A and (0) are complementary summands of each other. Observation 1 provides the following useful strategies for verifying that certain submodules are direct summands. If $\alpha{:}A \to B$ is a monomorphism and we wish to show that $\text{im}(\alpha)$ is a direct summand of B, it is sufficient to find a map $\beta{:}B \to A$ for which $a\alpha\beta = a$ for all a in A. If $\alpha{:}A \to B$ is an epimorphism and we wish to show that $\text{ker}(\alpha)$ is a direct summand of A, then it is sufficient fo find a map $\beta{:}B \to A$ for which $b\beta\alpha = b$ for all b in B. Finally, if A is a submodule of B and we wish to show that it is a direct summand, it is sufficient to find a map $\beta{:}B \to A$ for which $a\beta = a$ for all a in A.

A direct summand B of a left R–module A may have many distinct complementary summands, as Exercises 1 and 2 show. There is a one–to–one correspondence between the set of

complementary summands of B and the set of splitting maps
for the inclusion map of B into A. The complementary
summands for the summand $im(\lambda_B)$ of the coproduct A II B
are precisely the graphs of the homomorphisms of A into B.

EXERCISES

1. Let A be a noncyclic $\mathbb{Z}$–module of order 4. Show that
A is the direct sum of any two of its three cyclic submodules
of order 2.

2. Let R be the ring of integers modulo 2. Show that each
submodule of order 2^n of each R–module of order 2^{n+1} is a
direct summand having precisely 2^n distinct complementary
summands.

3. Let B be a direct summand of a left R–module A and let
$\alpha{:}B \rightarrow A$ be the inclusion map. Show that the function
associating each splitting map $\beta{:}A \rightarrow B$ of α with $ker(\beta)$
is a one–to–one correspondence between the set of splitting
maps of α and the set of all complementary summands of B
in A.

4. Show that the complementary summands of the submodule
$im(\lambda_B)$ of the left R–module A II B are precisely the graphs
of the homomorphisms of A into B.

5. Let A be the left $\mathbb{Z}$–module $\mathbb{Z}/(2)$ II $\mathbb{Z}/(4)$. Let B =
$\{(0,0), (1,0)\}$ and let C = $\{(0,0), (1,2)\}$. Show that both B and
C are direct summands of A but that B $\oplus$ C is not.

§9 NATURAL MAPS

Let B be a submodule of a left R–module A and let
A/B be the associated factor module described in §2. Since

each element of A lies in precisely one coset of B, we may define a function ν:A $\to$ A/B by specifying that, for each a in A, aν is that coset to which a belongs. In the usual coset notation, this function is described by aν = a+B. Since (a+a')ν = (a+a')+B = (a+B) + (a'+B) = aν + a'ν and (ra)ν = ra+B = r(a+B) = r(aν) hold for all a and a' in A and all r in R, ν is an R–homomorphism. The map ν is called the *natural map* from A to A/B.

Using natural maps (and inclusion maps), we can analyze arbitrary maps in a revealing way. If α:A $\to$ B is a map of left R–modules then we can write it as a composition $\nu\alpha'\lambda$, where:

(1) ν:A $\to$ A/ker(α) is the natural map;
(2) λ:im(α) $\to$ B is the inclusion map;
(3) α':A/ker(α) $\to$ im(α) is the map defined by setting (a + ker(α))α' = aα.

This is in fact a well–defined function for if a + ker(α) = a' + ker(α) then a – a' $\in$ ker(α) and so 0 = (a – a')α = aα – a'α, showing that aα = a'α. The image of α' is clearly the same as the image of α. Moreover, if (a + ker(α))α' = 0 then aα = 0 so a $\in$ ker(α). Therefore α' is both an R–epimorphism and an R–monomorphism and so is an R–isomorphism. Thus we have verified the following:

OBSERVATION 1. Any map between R–modules can be written as the composition of a natural map, an isomorphism, and an inclusion map.

The above construction embodies a great deal of interesting information about the nature of homomorphisms. Here are some easily–quoted immediate consequences.

OBSERVATION 2. If α:A $\to$ B is an R–epimorphism then B is isomorphic to A/ker(α).

OBSERVATION 3. Every map can be expressed as an epimorphism followed by a monomorphism.

Perhaps the central significance of Observation 1 is that it reduces the problem of determining all maps from a left R–module A to a left R–module B to the following program: determine all of the submodules of A and B; for each pair of submodules A' of A and B' of B, determine all isomorphisms α' (if any) between A/A' and B'. Each triple (A',α',B') for which $\alpha':A/A' \to B'$ is an isomorphism then yields a map $\alpha:A \to B$ and every map from A to B arises in this fashion precisely once. As for the cardinality of the set of maps from A to B, this can be expressed in the especially simple form $\Sigma\{h(B')k(B') \mid B'$ is a submodule of B$\}$, where h(B') is the cardinality of the set of automorphisms of B' and k(B') is the cardinality (possibly 0) of the set of those submodules A' of A for which A/A' is isomorphic to B'.

EXERCISES

1. Let B and C be submodules of a left R–module A, and let $\nu:A \to A/C$ be the natural map. Let $\alpha:B + C \to A/C$ be the restriction of ν to B + C, and let $\beta:B \to A/C$ be the restriction of ν to B.
(a) Find the kernels and images of α and β.
(b) Show that $(B + C)/C \cong B/(B \cap C)$.

2. If B is a direct summand of a left R–module A and if C and C' are submodules of A satisfying $B \oplus C = A = B \oplus C'$, show that $C \cong A/B \cong C'$.

3. Let $\{A_i \mid i \in \Omega\}$ be a family of left R–modules and, for each i in Ω, let B_i be a submodule of A_i. Show that $(\amalg A_i)/(\amalg B_i) \cong \amalg(A_i/B_i)$.

§10   SUBSETS AND THE SUBMODULES THEY GENERATE

Let A be a left R–module. Given a single element a of A, what can we build? We can multiply a by each scalar r in R to produce the set $\{ra \mid r \in R\}$, which will be denoted by Ra. It is easily verifiable that Ra is a submodule of A and, since a = 1a, this submodule contains a. Since any submodule of A that contains a must contain Ra, we observe that for any element a of a left R–module A there exists a unique smallest submodule of A containing a, namely Ra. We say that Ra is the submodule of A *generated* by a (or by $\{a\}$). Given two elements a_1 and a_2 of A, what can we build? We can build Ra_1 and Ra_2, and from these submodules of A we can form the set of all sums $\{r_1a_1 + r_2a_2 \mid r_1, r_2 \in R\}$, which will be denoted by $Ra_1 + Ra_2$. Again, it is easily verifiable that $Ra_1 + Ra_2$ is a submodule of A that contains both a_1 and a_2. Since any submodule of A that contains both a_1 and a_2 must contain $Ra_1 + Ra_2$, we observe that there exists a smallest submodule of A containing $\{a_1, a_2\}$, and that $Ra_1 + Ra_2$ is this submodule. We will say that $Ra_1 + Ra_2$ is the submodule of A generated by $\{a_1, a_2\}$.

The above discussion can be easily extended to any finite subset $\{a_1, ..., a_n\}$ of A. The submodule of A generated by this subset is $Ra_1 + ... + Ra_n = \{r_1a_1 + ... + r_na_n \mid r_1, ..., r_n \in R\}$, which is the unique smallest submodule of A containing $\{a_1, ..., a_n\}$. Now consider an arbitrary nonempty subset U of a left R–module A. The submodule of A *generated* by U is $\{\Sigma r_u u \mid u \in U$ and $r_u = 0$ for all but finitely–many values of u$\}$, which is the unique smallest submodule of A containing U. By convention, we define the submodule of A generated by ϕ to be (0).

Let A be a left R–module. We say that A is *cyclic* if there is an element a of A for which A = Ra. Likewise, A is *finitely–generated* if there is a finite subset $\{a_1, ...,$

a_n} of A for which $A = Ra_1 + ... + Ra_n$ and A is
countably-generated if either A is finitely-generated or
there exists a countably-infinite set $U = \{a_1, a_2, ...\}$ of
elements of A such that A is generated by U.

For a first set of examples, let A be a three-dimensional
vector space over the field $\mathbb{R}$ of real numbers, which we
identify with three-dimensional Euclidean space. If $u = 0$
then the submodule (that is, the subspace) generated by u is
(0); if $u \neq 0$ then the submodule generated by u is the line
containing u and the origin. The submodule generated by $\{u_1,$
$u_2\}$ is $\mathbb{R}u_1 + \mathbb{R}u_2$, which is (1) (0) if $u_1 = 0 = u_2$; (2) the
plane containg u_1, u_2, and the origin if these three points are
not collinear; (3) the line containing u_1, u_2, and the origin of
these three points are collinear. Every submodule of A is
finitely-generated (by three or fewer elements).

For a second set of examples, we will let A be the
additive group (that is, $\mathbb{Z}$-module) of real numbers. The
submodule (that is, subgroup) generated by $\{1/2, 2/3\}$ is
$\mathbb{Z}(1/2) + \mathbb{Z}(2/3)$, which is actually a cyclic submodule since it
is identical with $\mathbb{Z}(1/6)$. The submodule generated by $\{1, \sqrt{2}\}$
is $\mathbb{Z} + \mathbb{Z}\sqrt{2} = \mathbb{Z} \oplus \mathbb{Z}\sqrt{2}$. The submodule $\mathbb{Q}$ consisting
of the rational numbers is not finitely-generated since for each
finite subset $\{n_1/d_1, ..., n_k/d_k\}$ of $\mathbb{Q}$ we have $\mathbb{Z}(n_1/d_1) + ... +$
$\mathbb{Z}(n_k/d_k) \subseteq \mathbb{Z}(1/d_1 \cdot ... \cdot d_k)$. One of the more interesting generating
sets for $\mathbb{Q}$ is $\{1/n! \mid n$ a positive integer\}; every infinite
subset of this generating set is also a generating set for $\mathbb{Q}$.

The approach we have given to submodules generated by
subsets might be thought of as an approach from the inside. An
equivalent approach from the outside can be based on the
important (and mentally verifiable) fact that the intersection of
any nonempty family of submodules of an R-module A is
again a submodule of A. Since the submodule of a left
R-module A generated by a subset U of A is the smallest
submodule of A containing U, we can make the following
observation.

OBSERVATION 1. If U is a subset of a left R-module
A then the submodule of A generated by U is the

intersection of all submodules of A containing U.

This outside approach is attractive but for working purposes the inside approach is usually more useful.

Cyclic modules will play an important role throughout our work. Finitely–generated modules receive close attention in Chapters 10, 11, and 13. Countably–generated modules are constantly under study in Part Three. Some attention to generating sets in general is given in Chapter 5.

EXERCISES

1. If $\{A_i \mid i \in \Omega\}$ is a family of submodules of a left R–module A, show that the submodule $\Sigma\{A_i \mid i \in \Omega\}$ of A defined in §6 is precisely the submodule generated by $\cup A_i$.

2. Show that a left R–module A is finitely generated if, for any family $\{A_i \mid i \in \Omega\}$ of submodules of A satisfying $\Sigma\{A_i \mid i \in \Omega\} = A$ there exists a finite subset Λ of Ω satisfying $\Sigma\{A_j \mid j \in \Lambda\} = A$.

3. Let A be a left R–module and let $\{A_i \mid i \in \Omega\}$ be a family of submodules of A that satisfies the following condition: $0 = \Sigma\{a_i \mid i \in \Lambda\}$ can hold for a finite subset Λ of Ω and elements $a_i \in A_i$ $(i \in \Lambda)$ only when $a_i = 0$ for all i. Show that the submodule B of A generated by $\cup\{A_i \mid i \in \Omega\}$ satisfies $B = \oplus\{A_i \mid i \in \Omega\}$.

§11 MAXIMAL NESTS OF SUBSETS

Let A be a set and let $\mathscr{F}$ be a family of subsets of A. A subfamily $\mathscr{N}$ of $\mathscr{F}$ is a *nest* in $\mathscr{F}$ if for every pair of sets $A \in \mathcal{N}$ and $B \in \mathcal{N}$ we have either $A \subseteq B$ or $B \subseteq A$. A nest $\mathcal{M}$ in $\mathcal{F}$ is a *maximal nest* in $\mathcal{F}$ if the only nest in $\mathcal{F}$ which contains $\mathcal{M}$ is $\mathcal{M}$ itself. In some families of subsets it is possible to describe one or more specific maximal nests. The situations in which we will use maximal nests are not usually of this sort. In fact, we will be assured of the existence of the maximal nests we need only by employing the following broad principle, which we will regard as an axiom of set theory:

MAXIMAL NEST PRINCIPLE: For each set A and each family $\mathcal{F}$ of subsets of A, there exists at least one maximal nest in $\mathcal{F}$.

We hope that the maximal nest principle will seem intuitively plausible and we believe that it should be used without self–consciousness. It is one of several widely–used set–theoretic principles equivalent to the Axiom of Choice––among them the Well–Ordering Principle and Zorn's Lemma. For further consideration of these principles and their role in the foundations of mathematics, you may wish to consult Halmos [1960] or Rubin and Rubin [1963, 1985].

EXERCISES

1. Let $\mathbb{Q}$ be the set of rational numbers and let $\mathcal{F}$ be the family of all finite subsets of $\mathbb{Q}$. Describe all of the maximal nests $\mathcal{M}$ in $\mathcal{F}$. State precisely which subsets of $\mathbb{Q}$ are of the form $\cup \mathcal{M}$ for some maximal nest $\mathcal{M}$ in $\mathcal{F}$.

2. Let $\mathbb{Q}$ be the set of rational numbers. Let $\mathcal{F}$ be the family of all nonempty proper subsets of $\mathbb{Q}$. For each x in $\mathbb{Q}$, let $U_x = \{y \in \mathbb{Q} \mid x < y\}$ and $C_x = \{y \in \mathbb{Q} \mid x \leq y\}$. Let $\mathcal{M}$ be the family consisting of the sets U_x for all x in $\mathbb{Q}$ and the sets C_x for all x in $\mathbb{Q}$. Is $\mathcal{M}$ a maximal nest

in $\mathscr{F}$?

3. In Exercise 2, replace the set $\mathbb{Q}$ of rational numbers by the set $\mathbb{R}$ of real numbers. Is $\mathscr{M}$ a maximal nest in this case?

1

The Ring as a Module over Itself

§1 INTRODUCTION

The backbone of this book is a sequence of theorems that describe the structure of certain classes of modules. Each of these theorems explicates how the modules in question may be constructed by means of one of at most four processes: (1) choosing a submodule of R, (2) forming a factor module of R, (3) building certain extensions of these factor modules (to be explained in Chapter 7), and (4) forming coproducts. Thus R, regarded as a left module over itself, is the foundation for this book. Since this module is so important, we will go carefully through its definition again. For the underlying additive abelian group we use R with its ring addition. For the scalar multiplication of the ring R on the abelian group R, we use the ring multiplication. Reviewing the axioms of a left R-module, we have the following:

(1) $r(r' + r'') = rr' + rr''$ by the left distributivity of ring multiplication;

(2) $(r + r')r'' = rr'' + r'r''$ by the right distributivity of ring multiplication;

(3) $(rr')r'' = r(r'r'')$ by the associativity of ring multiplication;
(4) $1r = r$ by the definition of the multiplicative identity of a
     ring.

In §2 we will discuss the submodules and factor modules
of R, which are the basic building blocks in module theory. The
multiplicative identity element of R is a fundamental tool in
the theory of R–modules. In §3 we will show how this identity
element can be used to produce strong results about mapping
properties related to R. In the process, we will make our first
acquaintance with the concept of a projective module. This
concept of projectivity is one of the most fundamental general
concepts in module theory––perhaps the most fundamental.

§2 MODULES ASSOCIATED WITH R

We will examine the submodules, factor modules, and direct
summands of the ring R regarded as a left module over itself.

Submodules of R. A subset of a ring R is a
submodule of the left R–module R if and only if it is an
additive subgroup of R and is closed under scalar
multiplication, which is the multiplication on the left by elements
of R. Thus, the conditions under which a subset of a ring is a
left submodule are precisely those under which it is a left ideal,
so we conclude that the submodules of the left R–module R
are its left ideals. Similarly, the submodules of the right
R–module R are its right ideals.

Factor modules of R. As a left R–module, R is
generated by the identity element 1. Each factor R/I, where
I is an arbitrary submodule (i.e., left ideal) of R, is generated
by the coset 1+I. This property of being cyclic is characterizing
for factor modules of R: a left R–module is isomorphic to a

factor module of R if and only if it is cyclic. This fact may be seen in the larger context of Observation 1, which will prove to be a basic tool in the study of modules.

OBSERVATION 1. For each element a of a left R—module A, there is one and only one map $\alpha:R \to A$ satisfying $1\alpha = a$. The image of α is the cyclic submodule of A generated by a.

VERIFICATION: Given an element a of a left R—module A, define a function α from R to A by $r\alpha = ra$ for all r in R. Then $(r+r')\alpha = (r+r')a = ra + r'a = r\alpha + r'\alpha$; $(rr')\alpha = (rr')a = r(r'a) = r(r'\alpha)$; and $1\alpha = 1a = a$. Thus α satisfies the requirements in Observation 1. Now let $\beta:R \to A$ be any map which satisfies $1\beta = a$. Then $r\beta = (r1)\beta = r(1\beta) = ra = r\alpha$ for each r in R and so $\alpha = \beta$. Thus α is the only map meeting the requirements of the observation. The image of α is $\{r\alpha \mid r \in R\} = \{ra \mid r \in R\} = Ra$, which is the cyclic submodule of A generated by a. $\square$

Two special types of cyclic modules will play basic roles in our studies—simple modules and uniform cyclic modules. We will introduce these and discuss them briefly.

DEFINITION. A left R—module is *simple* if and only if it is not (0) and has only two submodules: itself and (0).

A simple module is automatically cyclic. In fact, it is cyclic in a very strong way: each nonzero element a of a simple left R—module A generates the module. This is so since, if a is nonzero, Ra is not the zero module and so, by the simplicity of A, it must be all of A.

The presence of the multiplicative identity 1 of a ring forces the existence of at least one simple module. In proving this fact, we will also show that every such ring contains a maximal submodule (that is, left ideal) as stated in the following definition.

DEFINITION. A submodule B of a left R–module A is a *maximal* submodule if it is proper and there are no submodules of A properly contained in A and properly containing B.

OBSERVATION 2. Any proper left ideal I of a ring R is contained in at least one maximal left ideal H of R and, for each such H, R/H is a simple left R–module.

VERIFICATION: Consider the collection of all proper left ideals of R containing I. This collection must contain at least one maximal nest. Let $\{L_i \mid i \in \Omega\}$ be such a nest, and let H = $\cup L_i$. If 1 were in H, then 1 would also belong to one of the L_i, yielding $L_i \supseteq R1 = R$ and contradicting the supposition that each of the L_i is proper. Thus 1 is not in H and so H is a proper left ideal of R. Now suppose that K is a left ideal of R properly containing H. Then K properly contains each L_i and the collection of left ideals of R consisting of K and all of the L_i is a nest of left ideals properly containing $\{L_i \mid i \in \Omega\}$. The maximality of this latter nest implies that K = R. Thus H is a maximal left ideal of R.

Now let H be any maximal left ideal of R and let B be a submodule of R/H. Then K = $\{r \in R \mid r+H \in B\}$ is a left ideal of R containing H and contained in R. By the maximality of H, we either have K = H, in which case B is (0), or K = R, in which case B = R/H. Thus R/H is a simple module. □

Notice that in the verification of Observation 2 we made use of the Maximal Nest Principle, which is equivalent to the Axiom of Choice. In fact, Observation 2 is equivalent to the Axiom of Choice. Indeed, in Zermelo–Fraenkel set theory, the statement "Every unique factorization domain has a maximal ideal" is already equivalent to the Axiom of Choice. See Hodges [1979] for a proof.

DEFINITION. A left R–module is *indecomposable* if it

is not the direct sum of any pair of proper submodules. A left R-module is *uniform* if each of its submodules is indecomposable.

Stated another way, a left R-module is uniform if and only if each pair of nonzero submodules has a nonzero intersection. The $\mathbb{Z}$-module $\mathbb{Q}$ and each of its submodules are uniform. Any nonzero uniform $\mathbb{Q}$-module must be isomorphic to $\mathbb{Q}$. Uniform modules have also been termed *absolutely indecomposable* modules, which is a very descriptive name, especially in view of the following observations.

> **OBSERVATION** 3. Let B be a uniform submodule of a direct sum $A \oplus A'$, and let π and π' be the projections onto A and A' respectively. Then either the restriction of π to B or the restriction of π' to B is a monomorphism.

VERIFICATION: If neither of these projections were monic, we would have two nonzero modules of B, namely $\ker(\pi) \cap B$ and $\ker(\pi') \cap B$, which would have intersection (0). $\square$

The previous observation is not very surprising. On first contact, the next one is likely to be. Its verification will expose a feature of uniform modules that is a basic tool in their use. This feature may be informally described by saying that each nonzero element of a uniform left R-module is necessarily interrelated with the overall structure of the module. The situation is somewhat weaker than in the case of simple modules, in which, as we recall, a nonzero element had to be a generator.

> **OBSERVATION** 4. If B is a uniform submodule of A $= \oplus\{A_j \mid j \in \Omega\}$ then, for some element h of Ω, the restriction to B of the hth projection map π_h is a monomorphism.

VERIFICATION: If B is (0) then any element h of Ω will do, so we may suppose that B is nonzero. Choose a nonzero element b of B. Let Λ be the set of those indices j in Ω satisfying $b\pi_j \neq 0$. By definition of the direct sum, this is a finite set. Then $A = [\oplus\{A_j \mid j \in \Lambda\}] \oplus [\oplus\{A_j \mid j \in \Omega \setminus \Lambda\}]$. By the previous observation, the restriction to B of one of the two associated projections must be monic. It cannot be the projection onto the second summand, since that map sends b to 0. Thus the projection of B onto the first summand is monic. Let this projection be denoted by π, and let $B' = B\pi$. Then B' is a submodule of $\oplus\{A_j \mid j \in \Lambda\}$ and, since Λ is a finite set, we note that a finite number of applications of Observation 3 will yield an element h of Λ for which the projection of B' onto A_h is monic. The composite projection $B \to B' \to A_h$ is also monic, as required. $\square$

A *local ring* is a ring in which the set of noninvertible elements is closed under taking finite sums. In this case, the set of noninvertible elements of the ring is a two-sided ideal. Such rings are interesting and important in their own right, and have been extensively studied. We will look at modules over such rings in Chapter 13. For the moment, however, we will confine our attention to left R-modules A having the property that their endomorphism ring $\text{Hom}_R(A,A)$ is local. This condition is closely tied to the indecomposability of the module A.

OBSERVATION 5. A nonzero left R-module having a local endomorphism ring is indecomposable.

VERIFICATION: If A is a nonzero left R-module which is not indecomposable then we can write $A = A_1 \oplus A_2$, where each A_i is a proper submodule of A. For each i, let π_i be the projection of A onto A_i and let λ_i be the injection of A_i into A. Then $\pi_1\lambda_1$ and $\pi_2\lambda_2$ are both noninvertible endomorphisms of A. Their sum, however, is the identity map of A, which certainly is invertible. Thus the endomorphism ring of A is not local. ☐

Direct summands of the module R. The direct summands of the left R—module R are tied to certain special elements of R.

DEFINITION. An element e of a ring is an *idempotent* if and only if $e^2 = e$.

OBSERVATION 6. A submodule B of the left R—module R is a direct summand if and only if B = Re for some idempotent element e of R.

VERIFICATION: Suppose that R = B ⊕ B' and let 1 = e + e' be the associated decomposition of 1. Since e is in B and B is a left ideal of R, we have Re ⊆ B. On the other hand, if b is an arbitrary element of B then b = b1 = be + be', where be ∈ B and be' ∈ B'. Then be' = b − be ∈ B ∩ B' = (0) and so be' = 0. Therefore b = be ∈ Re and so we see that B = Re. Moreover, from the special case of b = e we obtain e = e^2 and so e is idempotent. Thus any direct summand of R is of the form Re, where e is an idempotent element of R.

Now suppose that e is an idempotent element of R. We will verify that R = Re ⊕ R(1−e) is a direct sum decomposition of R. For each element r in R, we have r = re + r(1−e), which shows that R = Re + R(1−e). But Re ∩ R(1−e) = (0) since, if r'e = r"(1−e) holds for elements r' and r" of R, then we have r'e = r'ee = r"(1−e)e = r"(e−ee) = 0. Thus R = Re ⊕ R(1−e). ☐

Because of their importance in structure theory, those modules that are isomorphic to direct summands of R have been given a special name.

DEFINITION. A *principal left R-module* is an R-module that is isomorphic to a direct summand of the left

R—module R.

To emphasize the significance of the classes of modules introduced thus far, we will indicate their function in our structural studies chapter by chapter. The fundamental building block used in Chapter 2 is the left R—module R itself. For Chapter 3 the basic building blocks are simple modules. In Chapter 4 the blocks will be the left ideals of R. The uniform cyclic modules are fundamental blocks used throughout Chapter 9: in the final section of Chapter 9 the building blocks are also principal modules. In §2 of Chapter 10 the building blocks are again uniform cyclic modules. The finitely—generated left ideals of R are the blocks used in §3 and §4 of Chapter 11 (those in §4 are principal modules). Chapter 13 returns to R itself as the basic building block.

Aside from some elaborate decoration that must be added in certain cases to uniform cyclic modules (to be explained in Chapter 7), we have all the basic building blocks that are used in the theorems in this book. Our principal construction tool will be the coproduct. To emphasize this we will introduce a a terminological convention: the prefix *semi*, when applied to a type of module (but not necessarily to a type of ring!) will mean "a direct sum of". Thus a *semisimple module* is a direct sum of simple submodules, a *semiprincipal module* is a direct sum of principal submodules, a *semicyclic module* is a direct sum of cyclic submodules, and a *semiuniform module* is a direct sum of uniform submodules.

EXERCISES

1. Regard the ring **Z** of integers as a module over itself and study it as follows:
(a) Determine the submodules of **Z**.
(b) Determine the factor modules of **Z**.
(c) Which submodules of **Z** are maximal?
(d) Determine the simple **Z**—modules.

(e) Which cyclic $\mathbb{Z}$–modules are indecomposable?
(f) Which cyclic $\mathbb{Z}$–modules are uniform?
(g) Which elements of $\mathbb{Z}$ are idempotent?
(h) Determine the principal $\mathbb{Z}$–modules.
(i) Which cyclic $\mathbb{Z}$–modules are semisimple?
(j) Which cyclic $\mathbb{Z}$–modules are semiuniform?

2. Regard the ring $\mathbb{Q}$ of rational numbers as a module over itself. Carry out Exercise 1, replacing $\mathbb{Z}$ with $\mathbb{Q}$.

3. For some third ring R of your choice, regard R as a module over itself and attempt to carry out Exercise 1, replacing $\mathbb{Z}$ with R. Here are some choices of R which you may wish to consider: (1) the ring $\mathbb{Z}/(n)$ of integers modulo some positive integer n; (2) the ring $\mathbb{Q}[X]$ of all polynomials over $\mathbb{Q}$ in an indeterminate X; (3) the full ring all 2×2 matrices over $\mathbb{Q}$.

4. Let B be a uniform submodule of a coproduct $A = A_1 \amalg \ldots \amalg A_n$ of left R–modules. Show that for at least one of the projections $\pi_i : A \to A_i$ the restriction of π_i to B is a monomorphism.

5. Let R be an integral domain. Show that, as a module over itself, R is uniform.

6. Let A be a left R–module which is not uniform. Construct left R–modules B and B' such that $B \oplus B'$ contains a submodule C that is isomorphic to A and for which neither the projection of C onto B nor the projection of C onto B' is a monomorphism.

§3 USING THE IDENTITY OF R (PROJECTIVITY)

We have already used the identity of R to discuss cyclic modules. There is another elementary use of the identity that leads to strong conclusions concerning maps out of (and also onto) R. In discussing this use of the identity, we will meet one of the basic themes of module theory—projectivity.

OBSERVATION 1. Let A be a left R—module and let $\alpha:A \to R$ be an epimorphism. For any element a of A satisfying $a\alpha = 1$, the restriction of α to the cyclic submodule Ra of A is an isomorphism from Ra to R. Furthermore, $A = Ra \oplus \ker(\alpha)$.

VERIFICATION: The computation $(ra)\alpha = r(a\alpha) = r1 = r$, which is valid for every element r of R, shows that the restriction of α to Ra is bijective and consequently Ra $\cap$ $\ker(\alpha) = (0)$. Thus we need only verify that $A = Ra + \ker(\alpha)$. Indeed, let a' be an arbitrary element of A and let $r = a'\alpha$. Then $a' = ra + (a' - ra)$, where $ra \in Ra$ and $(a' - ra) \in \ker(\alpha)$, since $(a' - ra)\alpha = a'\alpha - ra\alpha = r - r1 = 0$. $\square$

Observation 1 can be derived from the following, more sweeping, observation.

OBSERVATION 2. For any epimorphism of left R—modules $\alpha:A \to B$ and any map $\beta:R \to B$ there is a map $\theta:R \to A$ satisfying $\beta = \theta\alpha$.

VERIFICATION: If a is any element of A for which $a\alpha = 1\beta$ then define the R—homomorphism $\theta:R \to A$ by $r\theta = ra$ for all r in R. Then $r\theta\alpha = (ra)\alpha = r(a\alpha) = r(1\beta) = (r1)\beta = r\beta$ for all r in R and so $\beta = \theta\alpha$. $\square$

The map θ is called a *lifting* of β or, in more detail, a lifting of β through α. In this language the observation may be stated more concisely: maps having domain R can be lifted through epimorphisms. This property has proven to be so significant that it has been incorporated in the following definition.

DEFINITION. A left R–module P is *projective* if, for each epimorphism α:A $\rightarrow$ B and each map β:P $\rightarrow$ B there is a map θ:P $\rightarrow$ A satisfying $\beta = \theta\alpha$.

We have already seen that R is a projective module. Which factor modules of R are projective?

OBSERVATION 3. A cyclic left R–module is projective if and only if it is a principal module.

VERIFICATION: For any cyclic left R–module B = Rb we have an epimorphism α:R $\rightarrow$ B given by rα = rb. Suppose that B is projective and let β:B $\rightarrow$ B be the identity map. Then there exists a map θ:B $\rightarrow$ R for which $\theta\alpha = \beta$. As we saw in Chapter 0, θ is a splitting map for α (that is, R = ker(α) $\oplus$ im(θ)) and so B, which is isomorphic to im(θ), is a principal module.

Now suppose that P is a principal module. Then P is isomorphic to a direct summand of R, so we may as well assume that R = P $\oplus$ P'. Let α:A $\rightarrow$ B be an epimorphism and let β:P $\rightarrow$ B be a map. Define the function γ:R $\rightarrow$ B by (p + p')γ = pβ, for p in P and p' in P'. Then γ is a map. Since R is projective, there exists a map φ:R $\rightarrow$ A satisfying $\varphi\alpha = \gamma$. Let θ be the restriction of φ to θ. Since the restriction of γ to P is β, we have $\beta = \theta\alpha$, as required. $\square$

The following observation highlights an important property of projective modules, which we will use often.

OBSERVATION 4. If α:A $\rightarrow$ P is an epimorphism of R–modules and if P is projective then α splits.

VERIFICATION: To show that α splits, we must show that there exists a map β':P $\rightarrow$ A for which $\beta'\alpha$ is the identity map on P. If we let $\beta:P \to P$ be the identity map, then such a β' must exist by the projectivity of P. ☐

Rings having the property that all of their left ideals are projective will arise in the course of our investigations in Chapter 4.

EXERCISES

1. Let $\alpha:A \to \mathbb{Q}$ be a homomorphism of $\mathbb{Q}$-modules for which $\text{im}(\alpha) \neq (0)$.
(a) Is α necessarily surjective?
(b) Show that A contains a direct summand isomorphic to $\mathbb{Q}$.

2. Let $\alpha:A \to \mathbb{Z}$ be a homomorphism of $\mathbb{Z}$-modules for which $\text{im}(\alpha) \neq (0)$.
(a) Is α necessarily surjective?
(b) Show that A contains a direct summand isomorphic to $\mathbb{Z}$.

3. Let $R = \mathbb{Z}/(4)$ be the ring of integers modulo 4. Let $\alpha:A \to R$ be an R-module homomorphism for which $\text{im}(\alpha) \neq (0)$. Can you necessarily conclude that A contains a direct summand isomorphic to R?

4. Prove that for any proper left ideal L of a ring R there is a maximal left ideal of R containing L.

5. Let A be a simple left R-module. Prove that R contains a maximal left ideal L for which R/L is isomorphic to A.

6. Show that there exists a projective simple left R-module if and only if some maximal left ideal of R is a direct summand of R.

7. Show that every simple left R—module is projective if and only if every maximal left ideal of R is a direct summand of R.

8. Show that a left R—module P is projective if and only if for each epimorphism of left R—modules $\alpha{:}A \to P$ there exists a map $\beta{:}P \to A$ such that $\beta\alpha$ is the identity map on P.

9. An element a of a left R—module A is *unimodular* if Ra is a direct summand of A isomorphic to R. Show that a is unimodular if and only if every element of R is of the form $a\alpha$ for some map $\alpha{:}A \to R$.

Part One

PROJECTIVITY

2

Coproducts of Copies of R (Free Modules)

§1 INTRODUCTION

In Chapter 1 we examined the ring R regarded as a module over itself. We made an introductory survey of the summands, submodules, and factor modules of R. Through the use of the identity element of R, we encountered the concept of a projective module. The pattern of development of Chapter 1 and the concepts introduced there prove an introduction to the general theory of projectivity. The role played by the R–module R in Chapter 1 will now be taken over by the class of all coproducts of copies of the module R or, more generally, by the free left R–modules.

DEFINITION. An R–module is *free* if it is isomorphic to a coproduct of copies of the module R.

Chapter 2 is concerned with the free modules themselves. After it, we will take up successively the direct summands of free modules, the submodules of free modules, and the factor modules of free modules. In the course of extending our attention from free modules to these broader classes, we obtain a thorough introduction to the concept of projectivity. We will return to deeper structural investigations of projective modules in §5 of Chapter 9.

§2 BASES

The definition we have chosen for the concept of a free module is itself a structural definition of these modules. However, since free modules play such a fundamental role in module theory, we will also reformulate this definition in terms of sets of elements, and we will practice using this alternate, element—wise description.

For an arbitrary ring R and an arbitrary set Ω, let $\{R_i \mid i \in \Omega\}$ be a family of copies of the R—module R indexed by the set Ω. Thus $R_i = R$ for each i in Ω. Let $F = \amalg\{R_i \mid i \in \Omega\}$. Then the set F consists of those functions f from Ω to R satisfying the condition that $f(i) = 0$ for all but finitely—many elements i of Ω. With each i in Ω we associate the element b_i of F defined by the conditions that $b_i(j) = 1$ if j equals i and $b_i(j) = 0$ otherwise. Let $B = \{b_i \mid i \in \Omega\}$. Now suppose that f is an arbitrary element of F and, for each i in Ω, let $r_i = f(i)$. Then $r_i = 0$ for all but a finite number of i in Ω and consequently the sum $\Sigma\{r_ib_i \mid i \in \Omega\}$ is defined and is an element of F. Since, for each j in Ω, we have $(\Sigma r_ib_i)(j) = \Sigma r_ib_i(j) = r_j$, we see that $f = \Sigma r_ib_i$. This B is a set of generators of F. But B has a further property. Suppose that $\{s_i \mid i \in \Omega\}$ is any indexed set of elements of R satisfying the condition that $s_i = 0$ for all but a finite number of i in Ω and that $f = \Sigma s_ib_i$. Then for each j in Ω we have $f(j) = (\Sigma s_ib_i)(j) = \Sigma s_ib_i(j) = s_j$, and consequently $s_j = r_j$ for every j in Ω. This uniqueness of the representation of f as a linear combination of the elements of B shows that B is a basis for F in the following sense.

DEFINITION. A subset B of a left R—module A is a *basis* for A if for each element a of A there is a unique indexed set $\{r_b \mid b \in B\}$ of elements of R for which $r_b = 0$ for all but a finite number of b in B and $a = \Sigma\{r_b b \mid b \in B\}$.

Vector spaces over fields always possess bases but this is not the case for modules over arbitrary rings, even relatively nice ones. For example, the set $\mathbb{Z}/(3)$ of integers modulo 3, considered as a left $\mathbb{Z}$—module, clearly has no basis. The possession of a basis is invariant under isomorphism. If $\alpha:A \to A'$ is an isomorphism of left R—modules and if B is a basis of A, then $B\alpha = \{b\alpha \mid b \in B\}$ is a basis of A'. It therefore follows that every free left R—module has a basis. We will show that the possession of a basis characterizes the free modules. For this purpose, the following powerful map—building tool will be convenient.

OBSERVATION 1. Let A be a left R—module having a basis B, and let A' be an arbitrary left R—module. For each function $g:B \to A'$ there is a unique map $\alpha:A \to A'$ satisfying $b\alpha = g(b)$ for all b in B.

VERIFICATION: Each element a of A can be written uniquely in the form $\Sigma\{r_b b \mid b \in B\}$, where the r_b are elements of R. If there is a map $\alpha:A \to A'$ which agrees with g on B, then it must satisfy $a\alpha = (\Sigma r_b b)\alpha = \Sigma r_b g(b)$. This shows that there is at most one map $\alpha:A \to A'$ which agrees with g on B. It also tells us that to complete the verification of the observation we need only consider the function α from A to A' which sends a to $\Sigma r_b g(b)$ and verify that it is a map and that it agrees with g on B. For a $= \Sigma r_b b \in A$ and $r \in R$ we have $(ra)\alpha = (\Sigma rr_b b)\alpha = \Sigma rr_b g(b)$ $= r[\Sigma r_b g(b)] = r(a\alpha)$. For $a = \Sigma r_b b$ and $a' = \Sigma s_b b$ we have $(a + a')\alpha = [\Sigma(r_b + s_b)b]\alpha = \Sigma(r_b + s_b)g(b) = \Sigma r_b g(b) + \Sigma s_b g(b) =$ $a\alpha + a'\alpha$. Finally, for an element b' of B the representation $b' = \Sigma r_b b$ is simply the one given by $r_{b'} = 1$

and $r_b = 0$ if $b \neq b'$. Thus $b'\alpha = \Sigma r_b g(b) = r_{b'} g(b') = g(b')$. □

From the uniqueness condition in Observation 1 we see that, if A is a left R−module with basis B, to verify that maps $\alpha{:}A \to A'$ and $\beta{:}A \to A'$ are identical we need only show that they agree on B. It is now convenient to observe the following basic characterization of free modules.

OBSERVATION 2. A left R−module is free if and only if it contains a basis.

VERIFICATION: We have already seen that a free module must have a basis; suppose that A is a left R−module having a basis I. Use I to index a family $\{R_i \mid i \in I\}$ of copies of R and form $F = \amalg\{R_i \mid i \in I\}$. Let $B = \{b_i \mid i \in I\}$ be the basis of F as constructed in the initial discussion of this section. Let $g{:}I \to B$ be the bijection given by $g(i) = b_i$ for all i in I, and let g' be the inverse of g. There is a unique map $\alpha{:}A \to F$ that agrees with g on I, and there is a unique map $\alpha'{:}F \to A$ that agrees with g' on B. Then $\alpha\alpha'$ is a map from A to A that agrees with the identity function g'g on I and $\alpha'\alpha$ is a map from F to F that agrees with the identity function gg' on B. It follows that $\alpha\alpha'$ and $\alpha'\alpha$ are identity maps and that α and α' are a pair of inverse isomorphisms. □

This observation suggests that a routine way of showing that a module is free is to construct a basis of the module.
A free module may possess more than one basis. For the **Z**−module **Z**, $\{1\}$ is a basis and so is $\{-1\}$. For **Z** $\amalg$ **Z**, the pair of elements $\{(1,0), (n,1)\}$ is a basis for any n in **Z**. Must two bases of the same free module have the same cardinality? The answer may be surprising. If the module has an infinite basis, order prevails: if F is a free module having an infinite basis, then any two bases of F must have the same cardinality. This fact is a consequence of the following

observation, which contains extra information as well.

OBSERVATION 3. If F is a free left R–module having a basis B of infinite cardinality k then each subset A of F having cardinality h less than k is contained in a proper direct summand of F.

VERIFICATION: Each element a of A has a unique representation of the form $a = \Sigma_{b \in B} r_b b$ with the elements r_b from R. Let $B(a) = \{b \in B \mid r_b \neq 0\}$. Then each $B(a)$ is finite and the cardinality of the union $C = \cup\{B(a) \mid a \in A\}$ is finite if A is finite and is equal to h if A is infinite. In either case, it is less than k and we have $F = [\oplus\{Rb \mid b \in C\}] \oplus [\oplus\{Rb \mid b \in B \setminus C\}]$ and $A \subseteq \oplus\{Rb \mid b \in C\} \neq F$. $\square$

If a free module does not have an infinite basis, chaos may reign. Among the exercises, you are asked to show that there exists a ring R for which for every positive integer n there is a basis of the left R–module R that has precisely n elements. This, however, can happen only if the ring R is noncommutative. In Section §4 we will see a class of rings R such that if F is a free left R–module having a finite basis then all bases of F have the same cardinality. At later points in this book, we will show that for rings of particular types, any two bases of the same free module must have the same cardinality.

We will now make an important observation about free modules.

OBSERVATION 4. Free modules are projective.

VERIFICATION: Let F be a free left R–module and let $\alpha : A \to A'$ be an arbitrary epimorphism of left R–modules. If $\beta : F \to A'$ is a map then we must construct a map $\beta' : F \to A$ satisfying $\beta'\alpha = \beta$. We begin by choosing a basis B of F. Since α is surjective, for each b in B there exists there exists an element a_b of A for which $a_b\alpha = b\beta$. Define a function $g{:}B \rightarrow A$ by $g(b) = a_b$. Then there is a unique map $\beta'{:}F \rightarrow A$ that agrees with g on B. For each b in B, $b\beta'\alpha = (b\beta')\alpha = (g(b))\alpha = a_b\alpha = b\alpha$. Thus $\beta'\alpha$ and β agree on a basis of F and so we conclude that $\beta'\alpha = \beta$, as required. ☐

   We have been explicit in mentioning the need for the Axiom of Choice when it arises and so at this point we should point out that Observation 4 is another statement equivalent to the axiom of choice. In fact, Blass [1979] has shown that the statement "every free $\mathbb{Z}$–module is projective" is equivalent, given the Zermelo–Fraenkel axioms for set theory, to the Axiom of Choice.
   It must be apparent that the specification of a basis of a free R–module F is very closely related to the specification of an isomorphism between F and a coproduct of an indexed family of copies of R. We will close out the discussion of bases by spelling out this relationship. Suppose that $B = \{b_i \mid i \in \Omega\}$ is a basis of F. Then $F = \oplus\{Rb_i \mid i \in \Omega\}$ and for each i in Ω the map $\alpha_i{:}R \rightarrow Rb_i$ defined by $r\alpha_i = rb_i$ is an isomorphism. The family of isomorphisms $\{\alpha_i \mid i \in \Omega\}$ may be assembled into an isomorphism $\alpha{:}\mathrm{II}\{R_i \mid i \in \Omega\} \rightarrow \oplus\{Rb_i \mid i \in \Omega\} = F$, where each R_i is a copy of R. Conversely, if $\alpha{:}\mathrm{II}\{R_i \mid i \in \Omega\} \rightarrow F$ is an isomorphism from the coproduct of an indexed family of copies of R to F and if $B = \{b_i \mid i \in \Omega\}$ is the basis of $\mathrm{II}R_i$ constructed at the beginning of this section, then $\{b_i\alpha \mid i \in \Omega\}$ is a basis of F.
   It is now clear that in discussing the bases of a free left R–module F we have been discussing the various ways of forming isomorphisms between F and coproducts of copies of R. We have learned of the existence of a ring R that has, for each positive integer n, a basis consisting of n elements. Expressing this fact in terms of coproduct representations gives, for such a ring, the following disconcerting sequence of R–module isomorphisms: $R \cong R \, \mathrm{II} \, R \cong R \, \mathrm{II} \, R \, \mathrm{II} \, R \cong \ldots$.

EXERCISES

1. Is (0) free? If so, what is its basis?

2. Let $\alpha:A \to A'$ be an isomorphism of left R–modules and suppose that A has a basis B. Show that $\{b\alpha \mid b \in B\}$ is a basis for A'.

3. Let R be a ring. We will construct from R a new ring R' that will be an infinite–dimensional analog of the ring of $n \times n$ matrices over R. The elements of R' will be the so–called *row–finite* matrices, namely those infinite matrices $[r_{ij}]$ with i,j positive integers such that, for each i, $r_{ij} = 0$ for all but finitely–many j. The addition and multiplication in R' are entirely analogous to the usual operations with $n \times n$ matrices: $[r_{ij}] + [s_{ij}] = [t_{ij}]$ where $t_{ij} = r_{ij} + s_{ij}$ for all i and j; and $[r_{ij}][s_{ij}] = [u_{ij}]$ where $u_{ij} = \Sigma\{r_{ik}s_{kj} \mid k$ a positive integer}. Notice that the apparently infinite sum is actually well–defined because r_{ik} is nonzero for only finitely–many k.
(a) Verify that R' is a ring.
(b) Let b_1 be the matrix $[r_{ij}]$ for which $r_{ij} = 1$ when $j = 2(i-1)+1$ and $r_{ij} = 0$ otherwise. Let b_2 be the matrix for which $r_{ij} = 1$ when $j = 2(i-1)+2$ and $r_{ij} = 0$ otherwise. Show that $\{b_1, b_2\}$ is a basis of R'.
(c) For each positive integer n, specify a basis of R' that consists of n elements.
(d) Does R' have an infinite basis?

4. Let R be an integral domain which is not a field and let F be the field of fractions of R. Show that F is not free as a left R–module.

5. Let R' be a ring, H be a proper two–sided ideal of R', and let R be the ring R/H. For each left R'–module A, let HA be the R'–submodule of A generated by the subset $\{ha \mid h \in H$ and $a \in A\}$.

(a) Verify that the left R'-module A/HA is also a left R-module with respect to the scalar multiplication defined as follows: if $r \in R$ and $a \in A$ then $(r+H) \cdot (a+HA) = ra + HA$.

(b) Show that if A is a free left R'-module with basis $\{b_1, ..., b_n\}$ then A/HA is a free left R-module with basis $\{b_1+HA, ..., b_n+HA\}$.

6. Suppose that F is a free left R-module which is not finitely generated and let P be a direct summand of F. Show that $P \amalg F \cong F$.

7. Let A be the infinite abelian group $\mathbb{Q}/\mathbb{Z}$. Does any submodule of A have a basis as a left $\mathbb{Z}$-module?

§3 INDEPENDENT SETS

This section is devoted to a concept of independence that provides a fundamental tool used in each part of this book.

DEFINITION. A subset U of an R-module A is *independent* if it does not contain 0 and the subset of A generated by U is the direct sum of the cyclic submodules Ru generated by elements u of U. The subset U is *uniformly independent* if it is independent and Ru is uniform for each u in U; it is *simply independent* if it is independent and Ru is simple for each u in U.

The following observation gives an example of such sets.

OBSERVATION 1. Let $\{A_i \mid i \in \Omega\}$ be a set of mutually nonisomorphic simple submodules of a left R-module A and let a_i be a nonzero element of A_i for each i in Ω. Then $\{a_i \mid i \in \Omega\}$ is a simply–independent subset of A.

VERIFICATION: If the observation were false then it would be false for some finite set Ω and so it suffices to consider the case that Ω is finite. Therefore we can assume that $\Omega = \{1, ..., n\}$ and proceed by induction on n. If n = 1 the result is certainly true. Assume therefore that n > 1 and that it is true whenever we have sets of at most n–1 elements of the appropriate type. If $\{a_1, ..., a_n\}$ is not simply independent there exist elements $r_1, ..., r_n$ of R such that 0 $\neq r_1 a_1 = r_2 a_2 + ... + r_n a_n$ and so $A_1 = Rr_1 a_1 \subseteq B = A_2 \oplus ... \oplus A_n$, where the left–hand equality is due to the simplicity of A_i and the directness of the sum comes from the induction hypothesis. Therefore there exists an index h $(2 \leq h \leq n)$ such the projection π_h from B onto A_h restricts to a nonzero map on A_1. By the simplicity of A_1 and A_h, this map then gives an isomorphism between A_1 and A_h, contradicting the choice of the A_i and so verifying the observation. $\square$

OBSERVATION 2. (1) The union of a nest of independent subsets of a left R–module is independent. (2) If the members of the nest are uniformly independent, then the union is uniformly independent. (3) If the members of the nest are simply independent, then the union is simply independent.

VERIFICATION: Let $\{U_i \mid i \in \Omega\}$ be a nest of independent subsets of a left R–module A. Let $U = \cup\{U_i \mid i \in \Omega\}$ and let B be the submodule of A generated by U. We must show that B is the direct sum of the submodules $\{Ru \mid u \in U\}$. To do this, we let U' be any finite subset of U and $\{r_u \mid u \in U'\}$ be a family of elements of R for which $\Sigma\{r_u u \mid u \in U'\} = 0$. We must show that $r_u u = 0$ for each u in U'. Indeed, since U' is finite and since U is the union of a nest of sets $\{U_i\}$, there exists an index j in Ω such that $U' \subseteq U_j$. The independence of U_j then proves the required equalities: $r_u u = 0$ for all u in U'. This completes the verification of (1); furthermore, (2) and (3) both follow immediately. □

The concept of independence will always be employed by means of the following observation.

OBSERVATION 3. Every left R—module A contains (1) a maximal independent set, (2) a maximal uniformly independent set, and (3) a maximal simply independent set.

VERIFICATION: (1) Consider the collection of all those subsets of A that are independent. This collection is not empty, since the empty subset of A is surely independent. The collection contains at least one maximal nest. Let $\{U_i \mid i \in \Omega\}$ be such a maximal nest, and let $U = \cup\{U_i \mid i \in \Omega\}$. Then U is independent and is not properly contained in any other independent subset of A for if U' were an independent subset of A properly containing U then U' would properly contain each of the U_i, contradicting the maximality of the given nest. Thus we have shown that U is a maximal independent subset of A.
(2) Reread the proof of (1) inserting "uniformly" before each occurrence of the word "independent". (3) Insert the word "simply" before each occurrence of the word "independent" in the proof of (1). □

A basis of a free left R—module is a maximal independent set. A basis of a nonzero free left R—module is uniformly independent if and only if R is uniform; it is simply independent if and only if R is simple. Regard $\mathbb{Z}$ as a left $\mathbb{Z}$—module. The maximal independent subsets of $\mathbb{Z}$ are the singleton subsets $\{n\}$, where n is nonzero. The only maximal independent subsets of $\mathbb{Z}$ that are bases are $\{1\}$ and $\{-1\}$. Each independent subset of $\mathbb{Z}$ is uniformly independent. The only simply independent subset of $\mathbb{Z}$ is the empty set that generates the submodule (0).

EXERCISES

1. Let $\mathbb{Z}/(6)$ be the additive group of integers modulo 6, and regard it as a $\mathbb{Z}$–module.
(a)   Find a uniformly independent subset of $\mathbb{Z}/(6)$ that generates it. Is this set simply independent?
(b)   Find an independent subset of $\mathbb{Z}/(6)$ that is not uniformly independent. Is this set a basis of the module?

2. The ring $\mathbb{Z}/(6)$ may be regarded as a module over itself. Rework the previous exercise regarding $\mathbb{Z}/(6)$ as a module over itself rather than as a $\mathbb{Z}$–module.

3. Regard the ring $\mathbb{Z}/(12)$ as a module over itself.
(a)   Does a uniformly independent subset exist that generates this module?
(b)   Does every maximal (uniformly) independent subset generate this module?
(c)   Is there a simply independent subset that generates this module?

4. Let R be a principal ideal domain and let $\{a_1, ..., a_n\}$ be a basis for a free left R–module A. If $b_1 = \Sigma r_i a_i$ $(r_i \in R)$, show that $\{b_1\}$ can be completed to a basis $\{b_1, ..., b_n\}$ for A if and only if the greatest common divisor of the r_i is 1.

5. Let F be a free left R–module with countable basis $\{x_1, x_2, ... \}$ and let $\{r_1, r_2, ... \}$ be a sequence of elements of R. Show that the subset $\{x_1-r_1x_2, x_2-r_2x_3, ...\}$ of F is independent.

6. An *independence structure* defined on a nonempty set A is a collection $\mathcal{E}$ of subsets of A satisfying the following conditions:
(1) $\phi \in \mathcal{E}$;

(2) If U is a subset of a member U' of $\mathcal{E}$ then U also

belongs to $\mathcal{E}$;

(3) If U and U' are finite members of $\mathcal{E}$ with U having less elements than U' then there exists an element U'' of $\mathcal{E}$ such that $U \subseteq U'' \subseteq U \cup U'$;

(4) If U is a subset of A satisfying the condition that every finite subset of U belongs to $\mathcal{E}$ then U belongs to $\mathcal{E}$.

Is the family of all independent subsets of a left R-module an independence structure in this sense? What about the family of all uniformly independent sets? What about the family of all simply independent sets?

§4   DIVISION RINGS

We have developed some familiarity with free modules. If a ring R has the property that all of its left R-modules are free, then we have some familiarity with all left R-modules. It is therefore interesting to ask which rings have the property that all left R-modules are free. The answer to this question will pick out the following class of rings.

DEFINITION. A ring R is a *division ring* if for each element b of R and each nonzero element a of R there exists an element c of R satisfying $ca = b$.

A commutative division ring is just a field. However, not all division rings are fields. The classical example of a division ring which is not a field is the ring of *real quaternions* defined by Hamilton. This ring is isomorphic to the subring of the ring of all 2×2 matrices of complex numbers consisting of those matrices of the form $\left[\begin{smallmatrix} a+bi & c+di \\ -c+di & a-bi \end{smallmatrix}\right]$ in which a, b, c, d are arbitrary real numbers. The proof that this is in fact a division ring is left as an exercise. A surprising, but classical, result of Wedderburn asserts that every finite division ring is a field. (See Chapter 7 of [Herstein, 1964] for a proof.)

Modules over division rings are called *vector spaces*. Most of the major results of linear algebra for vector spaces over fields hold when the scalars are taken from an arbitrary division ring.

OBSERVATION 1. A division ring is simple when regarded as a left module over itself.

VERIFICATION: Let I be a nonzero left ideal of R and let a be a nonzero element of I. If b is an arbitrary element of R then there exists an element c of R such that $b = ca \in I$, and so $I = R$. $\square$

Thus, if R is a division ring then every cyclic left R-module is isomorphic to R and is therefore simple and also uniform. Thus, for modules over division rings, every independent subset is uniformly independent and simply independent.

OBSERVATION 2. A ring R has the property that every left R-module is free if and only if it is a division ring.

VERIFICATION: Assume that every left R-module is free and let A be a simple left R-module. Since A is free, it is isomorphic to a coproduct of copies of R but, since it is simple, it is indecomposable and so must be isomorphic to R. Thus R itself is simple as a left R-module and so the only nonzero left ideal of R is R itself. In particular, if a is a nonzero element of R then $Ra = R$ and so for each element b of R there exists an element c of R satisfying $ca = b$. Thus R is a division ring.

Now suppose that R is a division ring and that A is an arbitrary left R-module. Let U be a maximal independent subset of A. Since, for each u in U, Ru is a nonzero cyclic left R-module, we have $Ru \cong R$. Thus, to show that A is free we need only show that U generates A. For a in $A \setminus$

U, the set $U \cup \{a\}$ is not independent by the maximality of U and so there exist elements r and $\{r_u \mid u \in U\}$ of R, not all of which are zero, such that $ra + \Sigma\{r_u u \mid u \in U\} = 0$. Since U is independent, r is nonzero. Since R is a division ring, there exists an element c of R satisfying $cr = 1$. Then $a = cra = c(-\Sigma r_u u) = \Sigma(-c r_u)u$. Thus U generates A, and A is free. $\square$

Notice that what we in fact showed in the verification of Observation 2 is that a ring R is a division ring if and only if every nonzero cyclic left R–module is free.

OBSERVATION 3. Let R be a division ring and let A be a left R–module having a basis $\{a_1, ..., a_n\}$. If $b = r_1 a_1 + ... + r_n a_n$ and $r_1 \neq 0$ then $\{b, a_2, ..., a_n\}$ is also a basis of A.

OBSERVATION 4. If R is a division ring then any two bases of the same left R–module have the same cardinality.

VERIFICATION: By Observation 3 of §2.2 we need only consider left R–modules which do not have an infinite basis. If A is such a module having the empty set as a basis, then A $= (0)$ and A has no other bases, so the observation is valid. This provides a base for an induction argument. Indeed, assume that A has a basis $\{a_1, ..., a_n\}$ having n elements and assume that the observation has already been shown to be valid for all left R–modules having a basis of less than n elements. If $\{b_1, ..., b_t\}$ is another basis for A with $t < n$ then we are done by the induction hypothesis. Hence we can assume that $t \geq n$ and all we are left to show is that we in fact have equality. Write $b_1 = r_1 a_1 + ... + r_n a_n$. There is an index j such that r_j is nonzero and, after renumbering the indices if necessary, we can in fact assume that $r_1 \neq 0$. By Observation 3, $\{b_1, a_2, ..., a_n\}$ is also a basis of A. Let B be the submodule of A generated by $a_2, ..., a_n$ and let C be the submodule of A generated by $b_2, ..., b_t$. Then $A = Rb_1 \oplus B$

$= Rb_l \oplus C$, and consequently $B \cong A/Rb_l \cong C$. Thus B has a basis of $n-1$ elements and a basis of $t-1$ elements. We conclude therefore that $n-1 = t-1$ and so $n = t$, as required. $\square$

We now detour briefly and use this result to verify an observation we made in §2.

OBSERVATION 5. If R is a commutative ring and A is a free left R–module having a basis $\{b_1, ..., b_n\}$ then any basis for A has precisely n elements.

VERIFICATION: By Observation 2 of §1.2 we know that R has a maximal ideal H. Moreover, the factor ring R/H is in fact a field. By Observation 3 of §2 we know that A has no infinite bases. Suppose that A has a finite basis $\{c_1, ..., c_m\}$. Then we have an R–isomorphism α from the coproduct B of n copies of R to A which sends an n–tuple $(r_1, ..., r_n)$ to $\Sigma r_i b_i$ and an R–isomorphism β from the coproduct C of m copies of R to A which sends an m–tuple $(s_1, ..., s_m)$ to $\Sigma s_j c_j$. The composite map $\theta = \alpha\beta^{-1}$ is an R–isomorphism from B to C.

If $r \in H$ and $x \in B$ then $(rx)\theta = r(x\theta) \in HC$ and so θ induces a map θ' from B/HB to C/HC which sends an element $x+HB$ to $x\theta+HC$. Clearly this map is an R–epimorphism since θ is. If $(x+HB)\theta' = 0+HC$ then there exist elements $r_1 ,..., r_k$ of H and $y_1, ..., y_k$ of C satisfying $x\theta = \Sigma r_i y_i$. Since θ is an isomorphism, each y_i is of the form $x_i\theta$ for some element x_i of B so $x\theta = \Sigma r_i x_i\theta$. But θ is monic so $x = \Sigma r_i x_i \in HB$. Therefore θ' is an R–isomorphism. However, B/HB and C/HC are also left (R/H)–modules with scalar multiplication defined by $(r+H)(x+HB) = rx+HB$ and $(r+H)(y+HC) = ry+HC$ respectively and θ' is an (R/H)–isomorphism. This is in clear contradiction to Observation 4, unless we have $m = n$. $\square$

The scope of Observation 4 is greatly extended by

Observation 6.

> **OBSERVATION** 6. Let R be a ring that is a
> homomorphic image of a ring R'. If R has the property
> that any two bases of the same free left R–module have
> the same cardinality, then R' also has this property.

A relatively elementary survey of some of the basic
research in division rings can be found in Dauns [1982].

EXERCISES

1. Verify that the ring of real quaternions is in fact a divison
ring.

2. Show that a ring R is a division ring if and only if R \
{0} is a group under multiplication.

3. For which rings R is it true that each maximal independent
subset of each left R–module A is necessarily a basis of A?

4. Let R be a division ring and let A be a left R–module
possessing a basis $\{a_1, ..., a_n\}$. Show that if $b = r_1a_1 + ... +
r_na_n$ and $r_1 \neq 0$ then $\{b, a_2, ..., a_n\}$ is also a basis for A.
You may wish to generalize this observation.

5. If A is a simple left R–module, show that the ring
$Hom_R(A,A)$ is a divison ring.

6. Let R' be a division ring. Can you find a ring R and a
simple left R–module A for which R' is isomorphic to the
ring $Hom_R(A,A)$?

7. Let R be the ring of all matrices of the form $\left[\begin{smallmatrix} a & b \\ 0 & c \end{smallmatrix}\right]$, with a,
b, and c elements of the field of integers modulo 2, and let I
be the left ideal of R consisting of all such matrices satisfying
$b = c = 0$. Show that R/I is not a simple left R–module but

that $\text{Hom}_R(R/I,R/I)$ is a division ring.

8. List all simple **Z**–modules. For each such module A, what is its ring of endomorphisms?

9. Let p be a prime integer and let R be the ring of polynomials in an indeterminate X with coefficients in the field **Z**/(p) of integers modulo p. Describe the simple left R–modules and their rings of endomorphisms. Is every finite field of characteristic p isomorphic to one of these endomorphism rings?

10. Verify Observation 6.

11. Let A be a free left R–module having an endomorphism ring which is a division ring. Show that $A \cong R$.

NOTES FOR CHAPTER 2

Everett [1942] was the first to construct examples of free modules over a ring R having finite bases of different sizes. Our proof that all bases of a free module having an infinite basis must have the same cardinality is based on that of Peinado [1964]. Observation 5 of §4 is due to Leavitt [1964]. For further interesting work on this problem, see Cohn [1966] or the appendix to Section IIB in McDonald [1984].

The notion of an independence structure, which appears in the exercises of §2.3, is used in general combinatorics and matroid theory. See Welsh [1976] for more details.

3

Direct Summands of Free Modules
(Projective Modules)

§1 INTRODUCTION

In this chapter we will expand our attention from the class of free modules to the class of all direct summands of free modules. Our goal is to develop as much insight into the structure of these modules as possible. Only the first steps can be taken in this chapter. We will give a survey discussion of this problem after making an observation that will allow us to shorten the phrase "a module isomorphic to a direct summand of a free module" to a single familiar word.

OBSERVATION 1. A left R-module is isomorphic to a direct summand of a free module if and only if it is projective.

VERIFICATION: Let $F = C \oplus D$, where F is a free left R-module. Let $\alpha:A \to B$ be an epimorphism of left R-modules and let $\beta:C \to B$ be a map. We must construct a map $\beta':C \to A$ satisfying $\beta'\alpha = \beta$. To do this, we first define the map $\gamma:F \to B$ by $(c+d)\gamma = c\beta$, where c and d are in C and D respectively. Notice that β is just the restriction of γ to C. Since F is projective, there is a map $\gamma':F \to A$ satisfying $\gamma'\alpha = \gamma$. We now define β' to be the restriction of γ' to C. Then $\beta'\alpha = \beta$ follows by restricting domains to C.

Now let P be a projective left R—module. For a moment, ignore the algebraic structure on P and regard it only as a set. Then we can consider a family $\{R_p \mid p \in P\}$ of copies of R, and form the free R—module $F = \amalg\{R_p \mid p \in P\}$. As a basis for F we have $B = \{b_p \mid p \in P\}$ where, for p and q in P, $b_p(q) = 1$ if $q = p$ and $b_p(q) = 0$ otherwise. The function $f:B \to P$ defined by $f(b_p) = p$ extends uniquely to an R—homomorphism $\alpha:F \to P$. This map is an epimorphism since f itself is surjective. Since P is projective, we know that α splits, and consequently $F = \ker(\alpha) \oplus P'$ for a submodule P' isomorphic to P. $\square$

This observation has two immediate consequences, which will be used frequently and without reference.

COROLLARY 1. A direct summand of a projective module is projective.

COROLLARY 2. The coproduct of any family of projective modules is projective.

Also, another consequence of the observation is no less important.

COROLLARY 3. If p is a nonzero element of a projective left R—module P then there exists a map α from P to R satisfying $p\alpha \neq 0$.

VERIFICATION: Since P is projective, it is a direct summand of a free left R—module F. If we think of p as a nonzero element of F, then there is a projection $\pi{:}F \to R$ of F onto one of its direct summands satisfying $p\pi \neq 0$. Now take α to be the inclusion map $P \to F$ composed with π. □

We can use Observation 1 to provide easy examples of projective modules which are not free. For example, consider the ring $R = \mathbf{Z}/(15)$. Since 3 and 5 are coprime, it is easy to check that $R \cong (5)/(15) \amalg (3)/(15)$ and so $(5)/(15)$ and $(3)/(15)$ are both projective left R−modules. However, they cannot possibly be free since the number of elements in a free left R−module must surely be a multiple of 15. Refer also to Exercise 6.

In asking for structural insight into direct summands of free modules, we have initiated the investigation of the structure of projective modules. Let us make a survey of all projective R−modules that we might regard as immediately available from the R−module R. Of course, R itself is the simplest example, but in Chapter 1 we also examined the submodules, factor modules, and summands of R. Under what circumstances were these associated modules projective? A submodule of R is simply a left ideal, which may or may not be projective. A cyclic R−module is projective if and only if it is isomorphic to a direct summand of R. Finally, every summand of R is a projective left ideal. Our search for projectives among the modules associated with R has yielded only the projective left ideals of R. When we close this class under coproducts, we have what we consider to be the broadest class of examples of projective modules that are readily available from R itself−−that is, the coproducts of projective left ideals of R.

If we hope for very strong restrictions on the structure of the projective modules over a given ring R, then an optimistic wish would be for a theorem of the following form. For the ring R, an R−module is projective if and only if it is isomorphic to a coproduct of projective left ideals of R. Each of the three basic structure theorems for projective modules that will be given in this book justifies this optimistic wish for a particular class of rings: quasi–Frobenius rings (Chapter 9), semihereditary rings (Chapter 13), and local rings (Chapter 13). These structure theorems appear late in the book after the consideration of more basic topics. The stuctural results presented in this chapter arise from an investigation of modules over those rings R for which every R–module is projective.

EXERCISES

1. Verify Corollary 1.

2. Verify Corollary 2.

3. Prove that a uniform projective left R–module must be isomorphic to a left ideal of R.

4. Show that a left R–module P is projective if and only if there exists a subset $\{x_i \mid i \in \Omega\}$ of P and a subset $\{\alpha_i \mid i \in \Omega\}$ of $\text{Hom}_R(P,R)$ such that, for each x in P we have $x\alpha_i = 0$ for all but finitely–many values of i and $x = \Sigma(x\alpha_i)x_i$.

5. Let $a_1, \ldots, a_n$ be (not necessarily distinct) elements of a commutative ring R and let F be the coproduct of n copies of R, for some positive integer n. Show that there exists an invertible $n \times n$ matrix with entries in R and top row equal to $x = (a_1, \ldots, a_n)$ if and only if Rx is a free direct summand of F having basis $\{x\}$.

6. Let R_1 and R_2 be rings and let $R = R_1 \times R_2$. Show that each R_i is a projective left R–module which is not free.

§2 SEMISIMPLE RINGS

For which rings R is every R–module projective? We will show that the answer is the class of rings described as follows.

DEFINITION. A ring R is a *semisimple ring* if R is semisimple when regarded as a left R–module.

According to this definition, R is semisimple if it is of the form $\oplus\{H_i \mid i \in \Omega\}$, where each H_i is a simple left ideal. An elementary, but important, fact is that Ω is necessarily finite. Let $1 = \Sigma\{r_i \mid i \in \Omega\}$ be the representation for the identity element with respect to this direct decomposition of R. Then $r_i \neq 0$ for only a finite subset Λ of Ω. Thus $1 = \Sigma\{r_i \mid i \in \Lambda\}$, and for each r in R we have $r = r1 = r\Sigma\{r_i \mid i \in \Lambda\} = \Sigma\{rr_i \mid i \in \Lambda\} \in \oplus\{H_i \mid i \in \Lambda\}$. It follows that $R = \oplus\{H_i \mid i \in \Lambda\}$ and so $\Omega = \Lambda$. For the remainder of our discussion, when we express a semisimple ring R in the form $R = \oplus\{H_i \mid i \in \Lambda\}$, it will be assumed that the H_i are simple left ideals. Notice that those semisimple rings for which the set Λ is a singleton are precisely the division rings.

The key to the structure of modules over semisimple rings is the following observation.

OBSERVATION 1. For each left ideal L of a semisimple ring $R = \oplus\{H_i \mid i \in \Omega\}$ there is a subset Λ of Ω for which $R = L \oplus [\oplus\{H_i \mid i \in \Lambda\}]$.

VERIFICATION: We can assume that $\Omega = \{1, ..., n\}$. Define L_1 to be $L \oplus H_1$ if $L \cap H_1 = (0)$ and to be L if $L \cap H_1 = H_1$. Continue in this manner for each subscript: define L_j to be $L_{j-1} \oplus H_j$ if $L_{j-1} \cap H_j = (0)$ and to be L_{j-1} if $L_{j-1} \cap H_j = H_j$. Then $L_n = L \oplus [\oplus\{H_i \mid i \in \Lambda\}]$ for some subset Λ of Ω and L_n contains each H_j $(1 \leq j \leq n)$. Consequently, $R = L_n$. $\square$

COROLLARY: The following conditions hold for a semisimple ring $R = \oplus\{H_j \mid 1 \leq j \leq n\}$:

(1) Each left ideal of R is cyclic;

(2) Each cyclic left R-module is semisimple;

(3) Each nonzero uniform cyclic left R-module is simple;

(4) Each simple left R-module is isomorphic to one of the H_j;

(5) Each semisimple left R-module is projective.

By using maximal simply independent sets, we can now determine the structure of all modules over a semisimple ring. Notice that for modules over a semisimple ring R every uniformly-independent set is simply independent but that, unless R is a division ring, there are independent sets which are not simply independent.

THEOREM 1. If R is a semisimple ring then every left R-module is semisimple.

PROOF: Let $R = \oplus\{H_i \mid 1 \leq i \leq n\}$ be a semisimple ring, and let A be an arbitrary left R-module. Let M be a maximal simply independent subset of A, and let B be the semisimple submodule $\oplus\{Rm \mid m \in M\}$ of A. By the maximality of M, B must contain every simple submodule of A. Then B also contains every semisimple submodule of A and, since every cyclic left R-module is semisimple, B contains every element of A. Thus A = B is semisimple. □

When a semisimple ring R is expressed in the form $R = \oplus\{H_i \mid 1 \leq i \leq n\}$, every simple left R-module is isomorphic to one of the H_j. This suggests that we rearrange the H_i and reindex them with double subscripts so that we produce a representation $R = [H_{11} \oplus ... \oplus H_{1,k(1)}] \oplus [H_{21} \oplus ... \oplus H_{2,k(2)}] \oplus ... \oplus [H_{m1} \oplus ... \oplus H_{m,k(m)}]$, where H_{ij} is isomorphic to H_{tu} if and only if i = t. With this indexing we see that every simple left R-module is isomorphic with precisely one member of the list $H_{11}, H_{21}, ..., H_{m1}$. With any m-tuple $(w_1, ..., w_m)$ of cardinal numbers, associate the left R-module that is the direct sum of w_1 copies of H_{11}, w_2 copies of H_{21}, ..., and w_m copies of H_{m1}. By our structure theorem, each left R–module is isomorphic with the module associated with precisely one m–tuple of cardinal numbers.

We can now answer very conveniently the initial question of this section: for which rings R is every left R–module projective?

OBSERVATION 2. A ring R has the property that all of its [simple] left R–modules are projective if and only if it is semisimple.

VERIFICATION: If R is semisimple then each of its modules is semisimple by Theorem 1 and projective by the corollary to Observation 1. Now suppose that R is a ring satisfying the condition that each of its simple left R–modules is projective. Let A be a maximal simply independent subset of R. Then $M = \oplus\{Ra \mid a \in A\}$ is a semisimple submodule of R which, by the maximality condition on A, contains every simple submodule of R. We will show that R is semisimple by showing that R = M. Suppose that this is not the case. Then there is a maximal left ideal H of R containing M. Since R/H is simple, it must be projective, and so R = H ⊕ K for a submodule K of R isomorphic to R/H. But then K ⊆ M ⊆ H, which is a contradiction. Thus we conclude that M = R. □

The verification of the following important observation will be left as an exercise.

OBSERVATION 3. A ring R is semisimple if and only if each submodule B of each left R–module A is a direct summand of A.

The two remaining sections of Chapter 3 are discussions of examples of semisimple rings.

EXERCISES

1. (a) For which nonnegative integers n is the ring $\mathbb{Z}/(n)$ of all integers modulo n a semisimple ring?
(b) Describe the nature of all $\mathbb{Z}/(30)$−modules.
(c) Describe a $\mathbb{Z}/(150)$−module that is not projective.

2. If a ring R has a nonprojective module, must it have a nonprojective cyclic module? Must it have a nonprojective left ideal?

3. Verify Observation 3.

4. Prove that a ring R is semisimple if and only if each (maximal) left ideal of R is a direct summand of R.

5. Must a ring be semisimple if and only if each of its left ideals is a principal module?

6. For each rational number r, verify that the subset $\{\begin{bmatrix} a & ar \\ b & br \end{bmatrix} \mid$ a, b $\in \mathbb{Q}\}$ is a simple left ideal of the full ring of 2×2 matrices over $\mathbb{Q}$.

7. Let A be a semisimple left R−module.
(a) Is every maximal simply−independent subset of A a maximal independent subset of A?
(b) Is every independent subset of A simply−independent?
(c) Does every maximal (simply−)independent subset of A necessarily generate A?

8. Verify the following observations concerning the class of semisimple left R−modules for an arbitrary ring R:
(a) Each submodule of each semisimple left R−module is a direct summand.
(b) Each submodule of each semisimple left R−module is semisimple.
(c) Each factor module of each semisimple left R−module is semisimple.

9.   Alter the wording of the verification of Observation 3 of §2.2 to produce a verification of the following observation: if A is a semisimple left R–module having a maximal simply–independent subset M of infinite cardinality k, then each subset of N of A having cardinality h < k is contained in a proper direct summand of A.

10. Show that if a semisimple left R–module A has an infinite (maximal simply–) independent subset, then every pair of maximal simply–independent subsets of A have the same cardinality.

11. Alter the wording of the verification of Observation 4 of §2.4 to verify the following observation: two maximal simply–independent subsets of the same semisimple left R–module A have the same cardinality.

12. Let {C_i | i ∈ Ω} be a set of simple left R–modules no two of which are isomorphic. Let D be a left R–module for which ⊕{A_i | i ∈ Ω} = D = ⊕{B_i | i ∈ Ω} where, for each i in Ω, A_i and B_i are direct sums of (possibly empty) families of submodules of D each isomorphic to C_i. Prove that $A_i = B_i$ for each i in Ω.

§3   MATRIX RINGS

We will discuss in an informal manner some examples of rings of matrices that are semisimple. Let D be a division ring and let R be the full ring of all 2×2 matrices with entries in D. Let L_1 be the set of all matrices in R of the form $\begin{bmatrix} a & 0 \\ b & 0 \end{bmatrix}$ and let L_2 be the set of all matrices in R of the form $\begin{bmatrix} 0 & a \\ 0 & b \end{bmatrix}$. Then L_1 and L_2 are both left ideals of R and the function $\alpha{:}L_1 \to L_2$ which sends the matrix $\begin{bmatrix} a & 0 \\ b & 0 \end{bmatrix}$ to $\begin{bmatrix} 0 & a \\ 0 & b \end{bmatrix}$ is an isomorphism of left R–modules, as is easily verified. We have a decomposition $R = L_1 \oplus L_2$ of R as a direct sum of the left R–modules L_i. Moreover, each of the L_i is simple. If either a or b is not zero then $R[\begin{smallmatrix} a & 0 \\ b & 0 \end{smallmatrix}] = L_1$ since for any c and d in D we can find elements w, x, y, and z in D so that we have $[\begin{smallmatrix} w & x \\ y & z \end{smallmatrix}][\begin{smallmatrix} a & 0 \\ b & 0 \end{smallmatrix}] = [\begin{smallmatrix} c & 0 \\ d & 0 \end{smallmatrix}]$. We have seen therefore that R is semisimple.

This discussion has been carried out for the full ring of 2×2 matrices with entries in D, but similar steps can be carried out for the full ring of n×n matrices, where n is any positive integer. Thus we see the following.

OBSERVATION 1. For any division ring D and any positive integer n the full ring of all n×n matrices with entries in D is semisimple.

In more detail, the ring of all n×n matrices with entries in a division ring D is the direct sum of n mutually isomorphic simple left ideals.

Let us return to the special case of the full ring R of 2×2 matrices with entries in D. Since R is semisimple, we know that every left R–module is a direct sum of simple left modules and that each simple left R–module is isomorphic to L_1. The same is true regarding the full ring of n×n matrices with entries in D.

OBSERVATION 2. For any division ring D and any positive integer n, the full ring R of n×n matrices with entries in D has, up to isomorphism, only one simple left R–module L, and every left R–module is isomorphic to a coproduct of copies of L.

Now let R' be the full ring of 2×2 matrices with entries in a division ring D' and let R'' be the full ring of 3×3 matrices with entries in a division ring D''. Let $R = R' \times R''$ and turn this into a ring by defining $(r',r'') + (s',s'') = (r'+s',r''+s'')$

and $(r',r'')(s',s'') = (r's',r''s'')$ for all r' and s' in R' and all r'' and s'' in R''. As a left R–module, we can easily verify that $R = L_1 \oplus L_2 \oplus K_1 \oplus K_2 \oplus K_3$, where L_i is the left ideal of R' consisting of those matrices possibly having nonzero entries only in the ith column (for $i = 1,2$) and K_j is the left ideal of R'' consisting of those matrices possibly having nonzero entries only in the jth column (for $j = 1,2,3$). Thus R is semisimple and each simple left R–module is isomorphic either to L_1 or to K_1. This argument holds for any finite direct product of full matrix rings over division rings, and so we have the following result.

OBSERVATION 3. If $R_1, ..., R_k$ are distinct full rings of matrices over division rings then $R = R_1 \times ... \times R_k$ is a semisimple ring having k mutually nonisomorphic simple left R–modules, and every left R–module is a direct sum of copies of certain of these simple modules.

All of our assertions about direct products of full matrix rings have admitted very simple and straightforward verifications. The final observation we wish to note is a famous result due to Wedderburn, which is remarkable but considerably more tedious to demonstrate, and somewhat beyond the program of this book. It is essentially the converse of our previous observations.

OBSERVATION 4. Every semisimple ring is isomorphic to the direct product of a finite number of full rings of matrices over division rings.

EXERCISE

1. From the discussion in this section, what can be said about the structure of commutative semisimple rings?

§4 GROUP RINGS

There is an extremely important class of rings that are semisimple but for which their semisimplicity is not apparent. They are a particular class of what are called group rings. We will begin by defining the group ring of an arbitrary group G over an arbitrary ring R. Ignore (for a moment) the algebraic structure on G, and regard G only as a set. Use G to index a family $\{R_g \mid g \in G\}$ of copies of R, and form the free R-module $R[G] = \amalg\{R_g \mid g \in G\}$. Recall that the elements of this free module are the finitely-nonzero functions $f: G \to R$. We will define a ring structure on $R[G]$. For the additive group we use the R-module addition of $R[G]$. We define the product of elements f and f' in $R[G]$ to be the function $f \cdot f'$ satisfying $(f \cdot f')(g) = \Sigma\{f(g')f'(g'') \mid g'g'' = g\}$. Since both f and f' are finitely nonzero, only finitely-many of the products $f(g')f'(g'')$ are nonzero; consequently, each such sum is defined, and $f \cdot f'$ is also finitely nonzero. With a little patience, the ring axioms can be verified for $R[G]$. This ring $R[G]$ is called the *group ring* of G over the ring R.

There is a natural and useful way to embed both G and R in $R[G]$. We have the usual basis $\{b_g \mid g \in G\}$ of the free R-module $R[G]$, where each b_g is defined by $b_g(g') = 1$ if $g = g'$ and $b_g(g') = 0$ if $g \neq g'$. It is easy to see that B is closed under the multiplication of $R[G]$ and that the function $\eta: G \to R[G]$ defined by $\eta(g) = b_g$ is a group isomorphism from G to B. The identity element of the ring $R[G]$ is the basis element b_1, where 1 denotes the identity element of G. The function $\gamma: R \to R[G]$ defined by $\gamma(r) = rb_1$ is a ring isomorphism from R to the subring $\{rb_1 \mid r \in R\}$ of $R[G]$. We will use the homomorphisms η and γ to embed G and R in $R[G]$. By this we mean that we shall write for each b_g simply g and thereby consider that G is a subset of $R[G]$. For each rb_1 we shall simply write r and thereby consider that R is a subset of $R[G]$. Let us see what this change of notation does for an arbitrary element f of R[G]. There is a unique indexed family $\{r_g \mid g \in G\}$ of elements of R for which $f = \Sigma r_g b_g$. This expression can now be rewritten $f = \Sigma r_g b_g = \Sigma r_g(b_1 b_g) = \Sigma(r_g b_1) b_g = \Sigma r_g g$, where each r_g and each g are regarded as elements of R[G] and where each term $r_g g$ is regarded as a product formed with respect to the multiplication of R[G].

Since R is a subring of R[G], every R[G]-module may also be regarded as an R-module. If $\alpha:A \rightarrow B$ is an R-homomorphism between left R[G]-modules and $(ga)\alpha = g(a\alpha)$ holds for all g in G and all a in A, then α is an R[G]-homomorphism by the following calculation: $((\Sigma r_g g)a)\alpha = (\Sigma(r_g ga))\alpha = \Sigma(r_g(ga)\alpha) = \Sigma(r_g g(a\alpha)) = (\Sigma r_g g)(a\alpha)$. This fact and one more will be useful in the verification of Observation 1. Each element r in R commutes with each element g in G: $r \cdot g = (rb_1)b_g = rb_g = (rb_g)b_1 = b_g(rb_1) = g \cdot r$.

OBSERVATION 1. If G is a finite group of order n and if R is a division ring of characteristic either zero or a prime not dividing n, then R[G] is a semisimple ring, and consequently every R[G]-module is semisimple.

VERIFICATION: To show that a ring is semisimple it suffices to show that every submodule of each module is a direct summand. Thus, let A be a submodule of a left R[G]-module B and let $\lambda:A \rightarrow B$ be the inclusion map. We need to find a map $\delta:B \rightarrow A$ for which $\lambda\beta$ is the identity map on A, since we will then have $B = A \oplus \ker(\delta)$. Since R is a subring of R[G], B is also a left R-module and A is an R- submodule of B. Since R is a division ring, there is an R-homomorphism $\alpha:B \rightarrow A$ such that $\lambda\alpha$ is the identity map on A. There is an easy way to modify α to produce an R[G]-homomorphism: first observe that for every g in G, $g^{-1}(gb)\alpha$ is an element of A. Thus, for each g in G, we have a function $\alpha_g:B \rightarrow A$ defined for each b in B by $b\alpha_g$

$= g^{-1}(gb)\alpha$. Each α_g is an R–homomorphism. Summing these functions gives another R–homomorphism $\beta = \Sigma\{\alpha_g \mid g \in G\}$ from B to A. This β is actually an R[G]–homomorphism: for each g' in G and each b in B we have $(gb)\beta = \Sigma\{g^{-1}(g(g'b))\alpha \mid g \in G\} = g'\Sigma\{(gg')^{-1}((gg')b)\alpha \mid g \in G\} = g'\Sigma\{g''^{-1}(g''b)\alpha \mid g'' \in G\} = g'(b\beta)$. The second form of the last equality follows from the fact that, for a fixed element g' of the group G, as g runs through all of the elements of G without repetition, so does gg'. For a in A we have $a\lambda\beta = a\beta = na$, where n is the order of G. Since the characteristic of R is either zero or a prime not dividing n, n has an inverse n^{-1} in R. We now define δ to be $n^{-1}\beta$ to obtain the desired R[G]–homomorphism from B to A. $\square$

The correspondence between matrix representations of groups and modules over group rings is outlined in the exercises.

EXERCISES

1. Let F be a field, let $V = F^n$ be the usual vector space (that is, F–module) of ordered n–tuples of elements of F. Let S be the full ring of $n \times n$ matrices with entries from F. Let G be a group and let $\varphi:G \to S$ be a function such that $\varphi(g)$ is nonsingular for each g in G and for which $\varphi(gh) = \varphi(g)\varphi(h)$ for all g and h in G. Functions such as φ are called *group representations*. Define an operation of the group ring F[G] on V by means of $(\Sigma\{r_g g \mid g \in G\})\cdot v = \Sigma\{r_g\varphi(g)v \mid g \in G\}$. Verify that with this operation V becomes a left F[G]–module.

2. Let F be a field, let G be a group, and let V be a left F[G]–module. Since we regard F as a subset of F[G], V also possesses a vector space structure over F. Assume that V has a finite dimension n as a vector space and let $\{b_1, ..,$

b_n} be an ordered basis for V over F. Each g in G provides a function μ_g:V → V defined by $\mu_g(v)$ = gv.

(a) Verify that μ_g is an F–homomorphism. Let S be the full ring of n×n matrices over F and let φ:G → S be the function defined by specifying that, for each g in G, $\varphi(g)$ is the matrix of the linear mapping μ_g with respect to the basis {b_1, ..., b_n}.

(b) Verify that, for each g in G, $\varphi(g)$ is nonsingular.

(c) Verify that $\varphi(gh) = \varphi(g)\varphi(h)$ for all g and h in G and that therefore φ is a group representation in the sense of Exercise 1.

3. Let F, G, V, S, φ be as in Exercise 1 and regard V as a left F[G]–module as described in that exercise. Using the standard basis b_1 = (1,0,...,0), ..., b_n = (0,...,0,1), in V, a group representation φ':G → S is provided by Exercise 2. Verify that $\varphi' = \varphi$.

4. Let F, G, V, S, φ be as in Exercise 2. Placing φ in the role of the function φ in Exercise 1, an F[G]–module structrue is provided for V by that exercise. Show that this structure is identical with the originally–given F[G]–module stucture on V.

5. Let F, G, V, S be as in Exercise 1 and let both φ:G → S and φ':G → S be group representations in the sense of that exercise. These representation are said to be *equivalent* if there exists a nonsingular matrix T in S for which $\varphi(g)$ = $T\varphi'(g)T^{-1}$ for all g in G. In accord with Exercise 1, both φ and φ' provide F[G]–module structures for V. Show that φ and φ' are equivalent if and only if the F[G]–modules thus defined are isomorphic.

6. If G is an abelian group and R is a commutative ring, show that the ring R[G] is commutative.

7. For any group G, show that a projective $\mathbb{Z}[G]$-module is a free abelian group.

8. Let F be a field and let G be a group satisfying the condition that $F[H]$ is a semisimple ring for all finitely-generated subgroups H of G. Show that $F[G]$ is semisimple as well.

NOTES FOR CHAPTER 3

The structural description of semisimple rings in terms of matrices over division rings that is discussed in §3 is one of the major classical results of Wedderburn. We do not need a proof of this theorem for our work but, to the student who would like to have one, we recommend the beautiful lecture by Emil Artin, "The influence of J. H. M. Wedderburn on the Development of Modern Algebra" [1950]. Other proofs are available in Burrow [1965] and Lambek [1966].

In §4 we have included the minimal amount of information needed to indicate how group-representation theory can be subsumed under the general structure theory of modules. The representations of a group G in terms of matrices over a field F need only be studied "up to equivalence." Thus, as we see from the exercises, the study of representations of G coincides with the study of the structure of $F[G]$-modules "up to isomorphism". In the most important classical situations, G is a finite group and F is a subfield of the field of complex numbers. In such situations Observation 1 of §4 applies and tells us that in the final analysis we need only study the nonisomorphic simple left ideals of the semisimple ring $F[G]$. These will be finite in number and will yield all of the so-called "inequivalent irreducible representations" from which all representations can be assembled. This observation is a very important result, due to H. Maschke [1898].

If you would like to see group representation theory discussed in the context of module theory, consult Curtis & Reiner [1962], Burrow [1965], Jacobson [1980], or Feit [1982]. For further information on the structure of group rings, see Passman [1977].

4

Submodules of Free Modules

§1   INTRODUCTION

We have made a preliminary investigation of summands of free modules.   What can we say about submodules of free modules?   We have one elementary tool which can be used to analyze such submodules.

> **OBSERVATION** 1.   Let   D   be a submodule of the left R–module $A \oplus B$,   and suppose that $D\pi$ is projective, where π is the natural projection of $A \oplus B$ onto   B. Then $D = [D \cap A] \oplus B'$ for some submodule   B'   of   B isomorphic to $D\pi$.

> **VERIFICATION:**   Let π' be the restriction of π to   D. Since $\mathrm{im}(\pi') = D\pi$ is projective, we know by Observation 4 of §1.3 that π' splits and therefore that $D = \ker(\pi') \oplus B'$ for some submodule   B'   of   B   isomorphic to $D\pi$.   Since $\ker(\pi') = D \cap A$,   we are done.   □

The manner in which projectivity is involved suggests that

the observation can only be used to analyze submodules of free modules over rings with the property that all left ideals are projective. These rings will be introduced in the next section.

§2 HEREDITARY RINGS

The broadest class of rings for which we will be able to describe all submodules of free modules is the class defined as follows.

DEFINITION: A ring R is *left hereditary* if each of its left ideals is projective.

All semisimple rings are left hereditary. The two classes of rings can be compared most readily through an alternate definition of semisimple rings: a ring R is semisimple if each of its left ideals is a direct summand. The ring $\mathbb{Z}$ of integers is hereditary since each nonzero ideal is of the form $n\mathbb{Z}$ and so is isomorphic to $\mathbb{Z}$ as a $\mathbb{Z}$-module. Similarly, the ring $D[X]$ of polynomials in an indeterminate X with coefficients in a division ring D can be shown to be hereditary. You will be asked to do this in the exercises.

We will approach the description of the submodules of free modules over hereditary rings very gradually by means of two observations.

OBSERVATION 1. Let R be a left hereditary ring and let A be a submodule of the free left R-module $F = R_0 \oplus \ldots \oplus R_{n-1}$, where each R_i is a submodule of F isomorphic to R. Then $A = H_0 \oplus \ldots \oplus H_{n-1}$, where each H_i is isomorphic to a left ideal of R.

This observation can be verified by induction on n. For n = 1, it holds for arbitrary rings and the required induction step is provided by Observation 1 of §4.1. We will not give the details since they are present in the verification of Observation 2, which may be interpreted to subsume Observation 1.

OBSERVATION 2. Let R be a left hereditary ring and let A be a submodule of the free left R–module $F = \oplus\{R_i \mid i \geq 0\}$, where each R_i is a submodule of F isomorphic to R. Then $A = \oplus\{H_i \mid i \geq 0\}$, where each H_i is isomorphic to a left ideal of R.

VERIFICATION: For each nonnegative integer j let $A_j = A \cap [\oplus\{R_i \mid 0 \leq i \leq j-1\}]$. Then $A_0 = (0)$ and by the nature of a direct sum (as opposed to a direct product), $A = \cup\{A_i \mid i \geq 0\}$. For each j let π_j be the projection of F onto R_j and let ρ_j be the restriction of π_j to A_{j+1}. For each j, $A_{j+1}\rho_j$ is a left ideal and is therefore projective. Since $\ker(\rho_j) = A_j$, we see that $A_{j+1} = A_j \oplus H_j$ for some $H_j \cong A_{j+1}\rho_j$. We note that $H_0 = A_1$ and observe that each H_i is isomorphic to a left ideal of R. We will show that $A = \oplus\{H_i \mid i \geq 0\}$. Indeed, if this sum were not direct, a contradiction could be produced as follows: if $0 = a_{i(1)} + \dots + a_{i(n)}$ with $i(1) < \dots < i(n)$, with $a_{i(j)} \in H_{i(j)}$, and with $a_{i(n)} \neq 0$, then the directness of $A_{i(n)+1} = A_{i(n)} \oplus H_{i(n)}$ would be contradicted. If $A \neq \oplus H_i$ then there would exist a least k for which A is not contained in $\oplus H_i$. Since $A_0 = (0)$, k could not be zero and the following contradiction of the choice of k would arise: $\oplus H_i \supseteq A_{k-1} \oplus H_{k-1} = A_k$. $\square$

Observations 1 and 2 show that a submodule of a free left module F over a left hereditary ring must be isomorphic with a finite direct sum of left ideals if F has a finite or countably–infinite basis. In §3 we will remove this cardinality restriction on F.

EXERCISES

1. Let L be a nonzero ideal of the ring $\mathbb{Z}$ of integers, and let n be a nonzero element of L having least absolute value.
(a) Verify that $L = \mathbb{Z}n$.
(b) Verify that the function $\alpha{:}\mathbb{Z} \to \mathbb{Z}n$ defined by $t\alpha = tn$ is an isomorphism of $\mathbb{Z}$–modules and consequently that $\mathbb{Z}$ is hereditary.

2. Let D be a division ring and let D[X] be the ring of polynomials in an indeterminate X with coefficients in D. Let L be a nonzero left ideal of D[X] and let g(X) be a nonzero element of L having least degree.
(a) Verify that $L = D[X]g(X)$.
(b) Verify that the function $\alpha{:}D[X] \to D[X]g(X)$ defined by $p(x)\alpha = p(X)g(X)$ is an isomorphism of left D[X]–modules and consequently that D[X] is left hereditary.

3. Prove Observation 1, using Observation 1 of §1 and finite induction.

4. If R is a ring then a projective left R–module A is said to be *hereditary* if each of its submodules is also projective.
(a) Modify the proof of Observation 2 to obtain the following generalization: let $\{A_i \mid i$ a nonnegative integer$\}$ be a family of hereditary left R–modules, and let B be a submodule of $\amalg A_i$. Then $B = \oplus B_i$, where each B_i is isomorphic to a submodule of the corresponding A_i.
(b) Show by example that, if the words "isomorphic to" were deleted from the statement in (a), the resulting sentence would be false.

5. Study the verification of Observation 2 to determine whether the submodules H_i of A are actually subsets of the corresponding submodules R_i of F.

6. Let $\mathbb{Z}$ be the ring of integers, and let F be the free

$\mathbb{Z}$-module $F = \mathrm{II}\{R_i \mid i$ a nonnegative integer$\}$, where each R_i is isomorphic to $\mathbb{Z}$. Let A be the submodule of F consisting of those functions f for which $f(0) = f(1)$ and $f(i) = 0$ for $i > 1$. The verification of Observation 2 provides submodules H_i of A for which $A = \oplus\{H_i \mid i$ a nonnegative integer$\}$. For each i, state specifically which elements of A lie in H_i.

§3 HEREDITARY RINGS (CONTINUED)

The key to the analysis of the submodules of free modules over hereditary rings is Observation 1 of §1. To make the analysis, we need only iterate the procedure given by this observation. Observations 1 and 2 of §2 were made by a finite iteration and a countably-infinite iteration respectively. What we need now is a vehicle that will allow iteration beyond the countable realm. For this purpose, we will use ordinal numbers. For an understanding of the ordinal numbers and their elementary properties, we suggest the book by Paul Halmos [1960]. The proof of Theorem 2 has been written to invite detailed comparison with the proof of Observation 2 of §2. The comparison will pinpoint the few specifically new features involved in iterating beyond countable realm. In the initial step of the proof of the theorem we will use the following equivalent of the well-ordering principle: for every set I there is an ordinal number m and a bijection between I and the set of all ordinal numbers less than m.

THEOREM 2. If R is a left hereditary ring and F is a free left R-module then every submodule of F is isomorphic to a coproduct of left ideals of R.

PROOF: If F is a free left R-module then there exists an ordinal number m such that F is isomorphic to $\amalg\{R_i \mid i <$ m}, where each R_i is a copy of R. (The free modules in Observations 1 and 2 of §2 have this form with m = n or m = ω, respectively.) Thus we can assume that $F = \oplus\{R_i \mid i < m\}$, where each R_i is a submodule of F isomorphic to R. Let A be a submodule of F. For each ordinal $h \leqq m$, let $A_h = A \cap [\oplus\{R_i \mid i < h\}]$. Then $A_0 = (0)$ and $A_m = A$. For each $h < m$, let π_h be the projection of F onto R_h and let ρ_h be the restriction of π_h to A_{h+1}. For each $h < m$, we see that $A_{h+1}\rho_h$ is a left ideal of R and is therefore projective. Since the kernel of ρ_h is A_h, there exists a submodule $H_h \cong A_{h+1}\rho_h$ of R such that $A_{h+1} = A_h \oplus H_h$. We note that $H_0 = A_1$ and observe that each H_i is isomorphic to a left ideal of R. We will show that $A = \oplus\{H_h \mid h < m\}$. If the sum were not direct, a contradiction could be produced as follows: if $0 = a_{i(1)} + ... + a_{i(n)}$ with $i(1) < ... < i(n) < m$, with $a_{i(k)} \in H_{i(k)}$ for all k, and with $a_{i(n)} \neq 0$, then the directness of $A_{i(n)+1} = A_{i(n)} \oplus H_{i(n)}$ would be contradicted. If $A \neq \oplus H_i$, then there would exist a least ordinal $t \leqq m$ for which A_t is not contained in $\oplus H_i$. Since $A_0 = (0)$, t could not be zero. If t has a predecessor, we have $A_{h-1} \subseteq \oplus H_i$ and $A_t = A_{t-1} \oplus H_{t-1} \subseteq \oplus H_i$. If t has no predecessor, then we have $A_t = \cup\{A_h \mid h < t\}$, and the inclusions $A_h \subseteq \oplus H_i$, which hold for all $h < t$, give $A_t \subseteq \oplus H_i$. Thus, a contradiction of the choice of t arises whether t has a predecessor or not. We conclude that $A = \oplus\{H_i \mid i < m\}$, as required. $\square$

This theorem has four corollaries. The first is a structure theorem for projective modules over left hereditary rings.

COROLLARY 1. If R is a left hereditary ring then every projective left R–module is isomorphic to a coproduct of left ideals of R.

The second and third corollaries are characterizations of left hereditary rings. The third explains the choice of the term "hereditary".

> **COROLLARY** 2. A ring R is left hereditary if and only
> if each submodule of each of the free left R–modules is
> projective.

> **COROLLARY** 3. A ring R is left hereditary if and only if
> each submodule of each projective left R–module "inherits"
> the projectivity of the containing module.

The final corollary provides an introduction to the next
section.

> **COROLLARY** 4. A ring R has the property that each
> submodule of each free left R–module is free if and only if
> each left ideal of R is free.

EXERCISES

1. Recall that a left R–module is hereditary if each of its
submodules is projective. Modify the statement and proof of
Theorem 2 to obtain a generalization concerning submodules of
direct sums of hereditary left R–modules.

2. Show that a coproduct of any family of hereditary left
R–modules is again hereditary. (In other words, show that the
semihereditary left R–modules are precisely the hereditary left
R–modules.)

§4 PRINCIPAL IDEAL DOMAINS

If we ask which *commutative* rings R have the property
that submodules of free R–modules are free, we arrive at one of

the most widely—used classes of rings.

DEFINITION. A commutative ring R is a *principal ideal domain* if it is an integral domain every ideal of which is cyclic.

The most common examples of principal ideal domains are the ring $\mathbb{Z}$ of integers and the ring F[X] of polynomials in one indeterminate X with coefficients in a field F. Another important example is the *ring of Gaussian integers* $\mathbb{Z}$[i] = {a + bi ∈ $\mathbb{C}$ | a, b integers}. Each field is also, trivially, a principal ideal domain.

OBSERVATION 1. The nonzero free ideals of a commutative ring R are precisely those ideals of the form Rr, where r is an element of R which is not a zero divisor.

VERIFICATION: Let F be a nonzero free ideal of a commutative ring R. Let r be an element of any basis of F. Then r is not a zero divisor since r'r = 0 and r' ≠ 0 would violate the defining property of the basis to which r belongs. If F has a basis consisting of two or more elements, say r and r', then we have F = Rr ⊕ Rr' ⊕ F', where F' is the submodule generated by the remaining elements of the basis to which r and r' belong. But then r'r = rr' ∈ Rr ∩ Rr' = (0), which yields rr' = 0, contradicting the fact that r cannot be a zero divisor. Thus F = Rr, where r is not a zero divisor.
    Let r be an element of R which is not a zero divisor. Then the kernel of the map α:R → Rr defined by r'α = r'r must be zero and so Rr ≅ R is free. □

OBSERVATION 2. A commutative ring R has the property that each submodule of each free R—module is free if and only if it is a principal ideal domain.

VERIFICATION: Suppose that R is a principal ideal domain. Then each nonzero ideal is of the form Rr, where r cannot be a zero divisior. Thus, every ideal of R is necessarily free, and by Corollary 4 of §3 submodules of free R-modules are necessarily free.

Suppose that R is a commutative ring for which submodules of free R-modules are necessarily free. Then every ideal of R is free; consequently, by Observation 1, every nonzero ideal of R is of the form Rr for some element r of R that is not a zero divisor. We need only show that R is an integral domain. Suppose u and v are nonzero elements of r for which uv = 0. Then Rv = Rr for some element r of R which is not a zero divisor and so there is an element w in R such that r = wv, and we have ur = uwv = 0, contradicting the assertion that r is not a zero divisor. We conclude that R is a principal ideal domain. □

Note that we have in fact proven a somewhat stronger observation, namely that a commutative ring R is a principal ideal domain if and only if each submodule of each finitely-generated left R-module is free.

We will close this section by mentioning a famous problem in ring theory. As we have already noted, if F is a field then the ring of polynomials F[X] is a principal ideal domain and so, by Observation 2, every projective left R-module is free. In 1955, Serre conjectured that the same condition holds for any ring $R = F[X_1, ..., X_n]$ of polynomials in n indeterminates over a field F. Since Bass, in 1963, had shown that non-finitely-generated projective modules over such rings are free, the conjecture reduced to the consideration of finitely-generated projective modules and stimulated considerable research activity, including the development of an entire field known as "algebraic K-theory". Finally, Serre's conjecture was settled in the affirmative in 1976 by D. Quillen and A. Souslin, working independently. For a survey of their results, as well as background to Serre's conjecture and its applications, see Lam [1978] or Rotman [1979]. Quillen and Souslin's results rely heavily on the commutativity of the field of coefficients of the polynomials. Ojanguren and Sridharan [1971], have given an example of a division ring R and a non–free projective $R[X_1,X_2]$–module.

EXERCISES

1. Let R be a principal ideal domain and let B be a submodule of a free left R–module A. Let $\{b_1, ..., b_m\}$ be a basis for B and let $\{a_1, ..., a_n\}$ be a basis for A. Show that $m \leq n$.

2. Let R be a principal ideal domain and let A be a free left R–module having a basis $\{a_1, ..., a_n\}$. If B is a submodule of A, show that we can find a basis $\{b_1, ..., b_n\}$ of A, a positive integer $m \leq n$, and elements $r_1, ..., r_n$ of R such that:
(a) r_i divides r_{i+1} for $1 \leq i \leq m$;
(b) $r_i = 0$ for $m < i \leq n$; and
(c) $\{r_1b_1, ..., r_mb_m\}$ is a basis for B.

3. Let $F[X]$ be the ring of polynomials in an indeterminate X over a field F and let A be a vector space over F. Let θ be an F–endomorphism of A. Define an operation of the ring $F[X]$ on A as follows: if $p(X) = s_0 + s_1X + ... + s_nX^n$ is an element of $F[X]$ and if a is an element of A then $p(X)a = s_0a + s_1(a\theta) + ... + s_n(a\theta^n)$. Verify that this operation provides the structure of a left $F[X]$–module for A.

4. Let $F[X]$ be the ring of polynomials in an indeterminate X over a field F, and let A be a left $F[X]$–module. Since F is a subring of $F[X]$, we know that A is also a left F–module. Let $\theta: A \to A$ be the function which sends an element a of A to Xa. Verify that θ is an F–endomorphism of A.

5. (a) Let A, F, and θ be as in Exercise 3, and regard A as a left F[X]-module as described in that exercise. From this F[X]-module structure, obtain an F-endomorphism θ' of A as in Exercise 4. Verify that $\theta' = \theta$.
(b) Let F, A, and θ be as in Exercise 4. Regard A as a vector space over F and use θ to provide A with an F[X]-module structure as in Exercise 3. Verify that this F[X]-module structure is identical with the originally-given F[X]-module structure on A.

6. Let F be a field and let A be a vector space over F. Endomorphisms θ_1 and θ_2 of A are said to be *similar* if there exists a nonsingular endomorphism ψ of A for which $\theta_1 = \psi\theta_2\psi^{-1}$. In the manner of Exercise 3, θ_1 provides an F[X]-module structure for A and so does θ_2. Show that θ_1 and θ_2 are similar if and only if these F[X]-module structures on A are isomorphic.

7. (a) Explain how Exercises 3 through 6 give precise meaning to the following assertion and how they justify this assertion: the problem of classifying linear transformations of vector spaces "up to similarity" is a special case of the problem of classfying modules over principal ideal domains "up to isomorphism".
(b) Let R be the full ring of n×n matrices over the field F. Matrices M and M' in R are *similar* if there is a nonsingular matrix N in R for which $M = NM'N^{-1}$. Explain how Exercises 3 through 6 justify the following assertion: the problem of classifying the matrices in R "up to similarity" is equivalent to the problem of classifying "up to isomorphism" the F[X]-modules that are n-dimensional when regarded as vector spaces over the subfield F of F[X].

NOTES FOR CHAPTER 4

One of the central topics in linear algebra is the determination of the similarity classes of linear transformations.

If we restrict our attention to linear transformations on finite-dimensional vector spaces and express the problem in matrix form, then we have the usual problem of finding the similarity classes of $n \times n$ matrices over a field. What is desired is then a collection of $n \times n$ matrices over the field that has the property that each $n \times n$ matrix is similar to exactly one matrix of the collection. The similarity theory of matrices leads to the rational canonical form and the Jordan canonical form for matrices. In the exercises to §4 we have included the minimal amount of information needed to indicate how the theory of similarity of linear transformations (and matrices) can be subsumed under the general structure theory of modules. If you would like to see this portion of linear algebra discussed in the context of module theory, we suggest Hartley & Hawkes [1970], Mac Lane and Birkhoff [1967], Rotman [1965], or Marcus [1978].

For more information on projective modules over principal ideal domains, refer to Auslander & Buchsbaum [1974].

Vector spaces over fields have the property that any basis for a subspace can be extended to a basis for the entire space. A commutative ring R is called a *Steinitz ring* if it has the analogous property, namely that any basis for a free submodule of a free left R-module can be extended to a basis for the entire module. These rings have been characterized by Chwe and Neggers [1970] and by Lenzing [1971].

5

Factor Modules of Free Modules

§1 INTRODUCTION

Throughout Part One we have expanded our attention successively from free modules to summands of free modules to submodules of free modules. In the present chapter we will discuss factor modules of free modules and of projective modules. This discussion represents the ultimate expansion of our attention since, as we shall see, every module is isomorphic to a factor module of a free module. This fact was implicit in the verification of Observation 1 of §3.1. We shall give a more thorough discussion here.

Let R be a ring and let A be a left R–module. We will survey the ways of representing A as a homomorphic image of a free module. The process can be organized into three choices: (1) choose a set G of generators of A; (2) choose a free module F having a basis B of cardinality greater than or equal to the cardinality of G; (3) choose a surjective function f:B → G and let α:F → A be the map determined by f. Notice that choice (2) is always possible since for every cardinal number there is a free module with a basis of that cardinal (see

§2.2). The map α is necessarily an epimorphism since its image contains the set G, which generates A. Notice too that every epimorphism $\alpha':F' \rightarrow$ A of a free module F' onto A may be regarded as having arisen from such choices: let B be a basis of F', let G be Bα', and let f be the restriction of α' to B. Then α' is the map corresponding to the choice of G as a generating set of A, F' is the free module having B as the appropriate basis, and f is the surjective function from B to G.

Which free modules possess A as a homomorphic image? To give the answer we associate a cardinal number k with A. Let k be the least cardinal number in the set {k' | k' is the cardinality of some set of generators of A}. Such a k exists since every set of cardinals contains a least element. From the previous paragraph we may conclude that A is a homomorphic image of the free module F if and only if F possesses a basis of cardinality at least k. In particular, each free module having a basis of cardinality k is, in a sense, a "smallest" free module having A as a homomorphic image.

EXERCISES

1. Let m and n be positive integers and consider the **Z**-module A = **Z**/(m) $\amalg$ **Z**/(n). What is the least cardinality of a set of generators of A?

2. Regard the additive group **Q** of rational numbers as a **Z**-module.
(a) What is the least cardinality of a set of generators for this module?
(b) Show that each set of generators for **Q** contains a proper subset which also generates **Q**.

3. Let A be a left R–module which is a homomorphic image of a finitely–generated free left R–module. Show that any subset G of A that generates A contains a finite subset

that generates A.

4. Let n be a positive integer. Find a cyclic $\mathbb{Z}$–module A that contains a generating set consisting of n elements and containing no proper subset that generates A.

5. Find a ring R having the property that each finitely–generated left R–module is cyclic.

6. Let A be a left R–module such that the least cardinality k of a set of generators for A is infinite, and let G be any subset of A that generates A. Show that G contains a subset of cardinality k that generates A.

§2 TRACES OF PROJECTIVE MODULES

If P and A are left R–modules then $\Sigma\{P\alpha \mid \alpha \in \text{Hom}_R(P,A)\}$ is a submodule of A, called the *trace* of P in A. In particular, the trace of P in R, denoted by tr(P), is a left ideal of R. If P is free, then tr(P) is all of R since among the maps from P to R is the projection $P \to R$ of P onto any of its direct summands, which, as we have seen, is an epimorphism. However, if P is not free then tr(P) need not be all of R. What can we say about tr(P)?

OBSERVATION 1. If P is a projective left R–module then tr(P) is a two–sided ideal of R.

VERIFICATION: We have already noted that tr(P) is a left ideal of R, and all that remains to show is that $r'r \in \text{tr}(P)$ whenever $r' \in \text{tr}(P)$ and $r \in R$. If $r' \in \text{tr}(P)$ then there exist elements $p_1, ..., p_n$ of P and maps $\alpha_1, ..., \alpha_n$ from P to R such that $r' = \Sigma p_i \alpha_i$. Moreover, for each i the function β_i from P to R which sends an element p to

$p\alpha_i r$ is also a map and so $r'r = \Sigma p_i \beta_i$, proving that $r'r$ belongs to $tr(P)$ as well. $\square$

OBSERVATION 2. If P is a projective left R-module then $tr(P)P = P$.

VERIFICATION: Since P is projective, there exists a left R-module P' satisfying the condition that $A = P \amalg P'$ is a free left R-module having basis B. In particular, each element p of P can be written as $\Sigma\{r_b b \mid b \in B\}$, where the r_b are elements of R. Let $\pi:A \to P$ be the projection onto the direct summand. Then $p = p\pi = \Sigma r_b b\pi$. On the other hand, the inclusion map $P \to A$ composed with the projection onto the bth summand $A \to R$ gives a map β_b from P to R defined by $p\beta_b = r_b$ for each b in B. Thus $p = \Sigma\{p\beta_b(b\pi)$ $\mid b \in B\} \in tr(P)P$. This shows that $P = tr(P)P$. $\square$

An important consequence of this observation is the following one, which we will need in §15.3.

OBSERVATION 3. If P is a projective left R-module then $tr(P)^2 = tr(P)$.

VERIFICATION: If $H = \text{Hom}_R(P,R)$ then, by Observation 2, we have $tr(P) = \Sigma\{P\alpha \mid \alpha \in H\} = \Sigma\{tr(P)P\alpha \mid \alpha \in H\} = tr(P)\Sigma\{P\alpha \mid \alpha \in H\} = tr(P)^2$. $\square$

What this observation means is that any element r of $tr(P)$ can be written as a finite sum $\Sigma s_i t_i$, where the s_i and t_i are elements of $tr(P)$.

We see that building traces defines a correspondence between the class of all projective left R-modules and a certain set of two-sided ideals of R. This correspondence is not one-to-one: all free left R-modules have the same trace. However, we can tell precisely when two projective left R-modules will correspond to the same ideal of R.

OBSERVATION 4. Projective left R—modules P and P' have the same trace in R if and only if each of them is a homomorphic image of a coproduct of copies of the other.

VERIFICATION: Assume that $tr(P) = tr(P')$. If β is a map from P to R and if p' is an element of P' then we have a map β' from P to P' defined by $p\beta' = (p\beta)p'$ for all p in P. The submodule of P' generated by the images of all maps of this type is $tr(P)P' = tr(P')P'$ and this, by Observation 2, is all of P'. On the other hand, this submodule is certainly contained in the trace of P in P' and so we see that the trace of P in P' is all of P'. Now let $H = Hom_R(P,P')$. For each α in H, let P_α be a copy of P. If $Y = \amalg\{P_\alpha \mid \alpha \in H\}$ then we can define a map $\beta:Y \to P'$ by $(\Sigma p_\alpha)\beta = \Sigma p_\alpha\alpha$. This function is well—defined since, by the nature of the coproduct, only finitely—many terms in the sum Σp_α are nonzero. The image of β is precisely the trace of P in P' and so P' is a homomorphic image of a coproduct of copies of P. Reversing the roles of P and P' in the above arguments, we see that P is also a homomorphic image of a coproduct of copies of P'.

Now assume that P and P' are homomorphic images of coproducts of copies of each other. Thus we have a coproduct $Y = \amalg\{P_i \mid i \in \Omega\}$ of copies of P and an epimorphism $\alpha:Y \to P'$. For each i in Ω let $\lambda_i:P \to Y$ be the embedding of P into the ith direct summand of Y. If β is a map from P' to R then $im(\beta)$ is the left ideal of R generated by $\{im(\lambda_i\alpha\beta) \mid i \in \Omega\}$ and so it is contained in $tr(P)$. Thus $tr(P') \subseteq tr(P)$. Reversing the roles of P and P' in this argument, we see that the reverse containment is also true and so $tr(P) = tr(P')$. $\square$

EXERCISES

1. Show that a left R—module P is projective if and only if

there exist a set $\{a_i \mid i \in \Omega\}$ of elements of R and a set $\{\alpha_i \mid i \in \Omega\}$ of maps from P to R satisfying

(a) if $p \in P$ then $p\alpha_i = 0$ for all but finitely–many values of i;

(b) if $p \in P$ then $p = \Sigma\{a_i p\alpha_i \mid i \in \Omega\}$.

(This is called the *dual basis property* of projective modules.)

2. Let R be a commutative ring and let P be a finitely–generated projective left R–module.

(a) Show that $P^* = \text{Hom}_R(P,R)$ is a finitely–generated projective left R–module.

(b) If M is another finitely–generated projective left R–module, show that so is $\text{Hom}_R(P,M)$.

3. Let P be a projective left R–module and let I be a two–sided ideal of R. Show that the following conditions are equivalent:

(a) $\text{tr}(P) \subseteq I$;

(b) $I \cdot \text{tr}(P) = \text{tr}(P)$;

(c) $IP = P$.

Part Two

INJECTIVITY

6

Enlarging Homomorphisms

§1   INTRODUCTION

A basic theme of Part One was the lifting of maps through epimorphisms.   The basic theme of Part Two is dual to the lifting problem.   Let   A   be a submodule of a left R–module B, and let $\alpha{:}A \to C$ be a map from   A to a left R–module   C. When can a map $\beta{:}B \to C$ be found that agrees with α on A?

We will begin by considering a more general question.   For an R–submodule A   of a left R–module   B   and for a map $\alpha{:}A \to C$,   under which circumstances can a submodule   A'   of B   and a map $\alpha'{:}A' \to C$ be found such that   A   is properly contained in   A'   and α' agrees with α on   A? The simplest result is stated in the following observation.

OBSERVATION 1.   If   A   and   B   are submodules of a left R–module   D and if $\alpha{:}A \to C$ and $\beta{:}B \to C$ are maps which agree on $A \cap B$ then they have a common enlargement $\gamma{:}A + B \to C$.

VERIFICATION. If $a + b$ is an element of $A + B$ we attempt a definition: $(a + b)\gamma = a\alpha + b\beta$. It is clear that γ will be a map with the desired properties if it is a well–defined function. If $a + b = a' + b'$ then $a - a' = b' - b \in A \cap B$ and $a\alpha - a'\alpha = (a - a')\alpha = (b' - b)\beta. = b'\beta - b\beta$ and so $(a + b)\gamma = a\alpha + b\beta = a'\alpha + b'\beta = (a' + b')\gamma$, which confirms that γ is well–defined. $\square$

Let us return to the problem of a submodule A of a left R–module B and a map $\alpha{:}A \to C$ and let us expand our attention to the set of all maps $\alpha'{:}A' \to C$ for which A' is a submodule of B containing A and α' agrees with α on A. The graph $gr(\alpha')$ of each such α' is a submodule of $B \amalg C$ containing $gr(\alpha)$ such that the restriction of the projection map $\pi{:}B \amalg C \to B$ to $gr(\alpha')$ is a monomorphism with image A'. Conversely, any submodule N of $B \amalg C$ that contains $gr(\alpha)$ and satisfies the condition that restriction of π to N is a monomorphism is the graph of a map from $N\pi$ to C which agrees with α on A. The family of submodules of $B \amalg C$ of the form $gr(\alpha')$, where α' is an enlargement of α, must contain a maximal nest. Let M be the union of such a maximal nest. Now $M = gr(\alpha')$ for some enlargement α' of α, according to the following considerations. Since M is the union of a nest of submodules containing $gr(\alpha)$, it is itself a submodule containing $gr(\alpha)$. If (b,c) and (b',c) are both elements of M then there exists a map β such that (b,c) and (b',c) are both elements of $gr(\beta)$. Since the restriction of π to $gr(\beta)$ is a monomorphism, $b = b'$. We conclude that the restriction of π to M is a monomorphism and $M = gr(\alpha')$, where the domain of α' is $A' = M\pi$ and where, for each x in A', $x\alpha'$ is that element y of C such that $(x,y) \in M$. Does there exist a map $\alpha''{:}A'' \to C$ for which A'' is a submodule of B properly containing A' and α'' agrees with α' on A'? There does not, because otherwise, by adjoining $gr(\alpha'')$ to our nest, we would produce a strictly larger nest, in contradiction to the assumed maximality. Thus we have verified the following observation.

> **OBSERVATION** 2. If A is a submodule of a left R–module B then for any R–homomorphism $\alpha:A \rightarrow C$ there is a submodule A' of B containing A and a map $\alpha':A' \rightarrow C$ which agrees with α on A and which cannot be enlarged to a map $\alpha'':A'' \rightarrow C$ for some submodule A'' of B properly containing A'.

We will refer to any such α' as a *maximal enlargement* of α.

In §2 we will shift our concentration from B and its submodules to C.

§2 THE INJECTIVITY CLASS OF A MODULE

Let R be a fixed ring and let C be a fixed left R–module. We examine the class of all left R–modules B having the property that if A is a submodule of B and $\alpha:A \rightarrow C$ is a map, then α can be extended to a map from B to C. The class of all such modules is the *injectivity class* of C and will be denoted by Inj(C).

> **OBSERVATION** 1. The injectivity class of a module is closed under taking factor modules.

VERIFICATION: Let C be a left R–module and let B/K be a factor module of a left R–module B in Inj(C). Any submodule of B/K is of the form A/K for some submodule A of B containing K. Let α be a map from such a submodule A/K to C, and let $\nu:A \rightarrow A/K$ be the natural map. Then there exists a map $\beta:B \rightarrow C$ extending $\nu\alpha$. Now define a map β' from B/K to C by $(b + K)\beta'$

$= b\beta$. To show that this map is well-defined, we note that if b + K = b' + K then $b - b' \in K$ so $(b - b')\beta = (b - b')\nu\alpha = 0$ so $b\beta = b'\beta$. Moreover, by definition we see that α is the restriction of β' to A/K.   □

OBSERVATION 2.   The injectivity class of a module is closed under taking coproducts.

VERIFICATION: Let C be a left R−module and let {B_i | i ∈ Ω} be a family of modules in Inj(C). Let A be a submodule of II{B_i | i ∈ Ω} and let α:A → C be an arbitrary map. Enlarge α maximally to a map β:M → C. We must show that M = II{B_i | i ∈ Ω}. Assume that this is not so. If j is an element of Ω then the restriction of β to M ∩ B_j can be extended to a map γ from B_j to C since B_j belongs to Inj(C). Since β and γ agree on the intersection of their domains, there is a map with domain M + B_j that enlarges β. But β cannot be strictly enlarged, and so we conclude that $B_j \subseteq$ M. This is true for any element j of Ω, and so II{B_i | i ∈ Ω} $\subseteq$ M, establishing equality.   □

Part Two of this book focuses on those left R−modules C for which every left R−module lies in Inj(C).

DEFINITION. An R−module C is *injective* if for each submodule A of each module B and each map α:A → C there is a map β:B → C which restricts on A to α.

The decision as to whether a given R−module C is or is not injective is generally made using the following result, known as *Baer's criterion*.

OBSERVATION 3.   A left R−module C is injective if and only if for each left ideal I of R and each map α:I → C there is a map β:R → C which restricts on I to α.

VERIFICATION: If C is injective this criterion is certainly satisfied. Conversely, the given criterion states that R belongs to Inj(C). Therefore, by Observations 1 and 2, factor modules of coproducts of copies of R belong to Inj(C). But in Chapter 5 we saw that any left R–module is of that form. Therefore C is injective. ☐

COROLLARY: A left R–module A is injective if and only if, for each projective left R–module B and each submodule B' of B, every map from B' to A can be enlarged to a map from B to A.

Baer's criterion can be used to show that specific modules are injective: for example, that $\mathbb{Q}$ is an injective left $\mathbb{Z}$–module. It can also be used to establish properties of the class of all injective modules. To illustrate this use of Baer's criterion, we will show the closure of the class of injective modules under direct summands.

OBSERVATION 4. Each direct summand of an injective left R–module is injective.

VERIFICATION: Let C' be a direct summand of an injective left R–module C. Let $\lambda{:}C' \to C$ be the inclusion map and let $\pi{:}C \to C'$ be the projection of C onto C'. If L is a left ideal of R and if $\alpha{:}L \to C'$ is a map then $\alpha\lambda$ is a map from L to C and so it can be enlarged to a map $\beta{:}R \to C$. Then $\beta\pi{:}R \to C'$ is a map which restricts to α on L. By Baer's criterion, C' is injective. ☐

Baer's criterion states that a left R–module C is injective if and only if, for any left ideal I of R, the homomorphism of abelian groups $\text{Hom}_R(R,C) \to \text{Hom}_R(I,C)$ which sends each map α to $\lambda\alpha$ (where $\lambda{:}I \to R$ is the inclusion map) is an epimorphism. We can try and dualize this result and ask a similar question for projective modules. If P is a left

R−module satisfying the condition that for every left ideal I of R the homomorphism of abelian groups $\text{Hom}_R(P,R) \rightarrow \text{Hom}_R(P,R/I)$ which sends each map α to $\alpha\nu$ (where $\nu{:}R \rightarrow R/I$ is the natural map) is an epimorphism, is P necessarily projective? If we consider the special case of $R = \mathbb{Z}$ and recall that projective left $\mathbb{Z}$−modules are free since $\mathbb{Z}$ is a principal ideal domain, this becomes a long−standing famous problem known as *Whitehead's problem*, the solution of which turns out to be very surprising. If the module P is countable, then K. Stein showed that our question has an affirmative answer. However, if P is uncountable then Shelah [1974] proved that the problem is undecidable! That is to say, both the statement "if P is uncountable and $\text{Hom}(P,\mathbb{Z}) \rightarrow \text{Hom}(P,\mathbb{Z}/(n))$ is an epimorphism for all integers n then P is free" and its negation are consistent with the usual axioms of set theory (namely, the Zermelo−Frankel axioms together with the Continuum Hypothesis). Thus we can adjoin a solution to Whitehead's problem−−either affirmative or negative as we wish−−to our axioms for set theory without obtaining a contradiction, a most unexpected state of affairs.

EXERCISES

1. If R is a commutative noetherian ring, show that an R−module C is injective if and only if every map from a prime ideal of R to C can be enlarged to a map from R to C.

2. If B is an injective submodule of a left R−module A, show that B is a direct summand of A.

3. Show that a ring R is semisimple if and only if all left R−modules are injective.

4. Show that a ring R is semisimple if and only if each of its (maximal) left ideals is injective.

5. Let $\varphi:R \to S$ be a homomorphism of rings. Then S has the structure of a left R–module if we define $r \cdot s = \varphi(r)s$ for all r in R and s in S. Moreover, as we have seen in the exercises to §0.5, $\text{Hom}_R(S,A)$ has the structure of a left S–module for every left R–module A. Show that this is an injective left S–module if A is an injective left R–module.

6. Let R and S be rings, let N be an injective right S–module, and let P be an (R,S)–bimodule which is projective as a left R–module. Show that $\text{Hom}_S(P,N)$ is an injective right R–module.

7. Let A be a left R–module having endomorphism ring S and let a be an element of A satisfying the condition that Ra is a simple submodule of an injective submodule of A. Show that aS is a simple submodule of A, considered as a right S–module.

§3 GENERATORS

In deriving Baer's criterion, the left R–module R was used as a generator module in the sense of the following definition.

DEFINITION. A left R–module G is a *generator* if every left R–module is an epimorphic image of a coproduct of copies of G.

The module R is not the only generator we could have chosen. For example, it should be immediately obvious that every free left R–module is also a generator. More generally, we have the following result.

OBSERVATION 1. A left R–module A is a generator if and only if $\text{tr}(A) = R$.

We may use this concept to restate Baer's criterion in a slightly stronger form: a left R—module C is injective if and only if Inj(C) contains a generator.

The concept of a generator has been used widely in recent years in several advanced branches of algebra. It may be interesting to locate this concept and some of its uses in Bass [1968], Mac Lane [1971], and Jacobson [1980]. The generators that are most widely used are also projective. They are sometimes referred to by the contracted form *progenerator*. For each ring R, any free left R—module is a progenerator.

EXERCISES

1. Let R be a semisimple ring. Which left R—modules are generators?

2. For which rings R is every nonzero left R—module a generator?

3. Which **Z**—modules are generators?

4. Prove that a left R—module G is a generator if and only if for every pair of left R—modules A, B and every pair of distinct maps $\alpha_1:A \to B$ and $\alpha_2:A \to B$ there is a map $\beta:G \to A$ for which $\beta\alpha_1$ and $\beta\alpha_2$ are distinct.

5. Prove that a left R—module G is a generator if and only if R is isomorphic to a direct summand of a direct sum of a finite number of copies of G.

6. Prove that a left R—module G is a generator if and only if R contains left ideals $L_1, ..., L_n$ and elements $x_i \in L_i$ (i = 1, ..., n) such that each L_i is a homomorphic image of G and $\Sigma x_i = 1$.

7. Let R be a ring with the property that each of its left ideals is free. Prove that a left R—module G is a generator if and only if G contains a direct summand isomorphic to R.

8. Verify Observation 1.

9. Let R be a commutative ring and let P be a finitely—generated projective left R—module satisfying the condition that for each nonzero element r of R there is an element x of P satisfying rx ≠ 0. Show that P is a progenerator.

NOTES FOR CHAPTER 6

The notion of an injective module is due to Reinhold Baer [1940], as is the criterion which bears his name. A compact proof of this result can also be found on page 9 of Cartan & Eilenberg [1956]. Several weaker versions of this result can be found in the literature, some of which are presented in the exercises. See Bican, Jambor, Kepka & Nemec [1976], Smith [1981] or Vámos [1983] to get a flavor of such results.

Needless to say, this criterion is very dependent on the fact that the ring R has a multiplicative identity element. The consequences of Baer's criterion for modules over a ring not having a multiplicative identity have been studied by Faith and Utumi [1964].

7

Embedding Modules in Injective Modules

§1   INTRODUCTION

It is one of the fortunate basic facts of module theory that for each ring R, every R-module can be embedded in an injective R-module. Baer's criterion indicates how we should begin the embedding process. If A is a left R-module which is not injective, then there is a left ideal L of R and a map $\alpha:L \to A$ that cannot be enlarged to a map from R to A. It is an easy matter to embed A in a larger module B so that α can be enlarged to a map from R to B. This embedding process, although simple, is the most fundamental construction in this chapter, and is described rather formally.

OBSERVATION 1. For a left R-module A, a left ideal L of R, and a map $\alpha:L \to A$, the graph $gr(-\alpha)$ of the map $-\alpha$ is a submodule of R II A. The function $\lambda:A \to (R\ II\ A)/gr(-\alpha)$ defined by $a\lambda = (0,a) + gr(-\alpha)$ is an R-monomorphism. The function $\beta:R \to (R\ II\ A)/gr(-\alpha)$ defined by $r\beta = (r,0) + gr(-\alpha)$ is a map that enlarges the composite $\alpha\lambda$.

VERIFICATION: There is little here that needs comment. The maps discussed satisfy the condition that $\alpha\lambda$ and $\iota\beta$ go from L to $(R \amalg A)/gr(-\alpha)$, where $\iota{:}L \to R$ is the inclusion map.   The last assertion of the observation is that these maps are equal, which we see as follows: if $x \in L$ then $x\beta = (x,0) + gr(-\alpha) = (x,0) + (x,-x\alpha) + gr(-\alpha) = (0,x\alpha) + gr(-\alpha) = x\iota\alpha$. $\square$

Referring to the observation, when a map $\alpha{:}L \to A$ is given, we may identify A with its image in $B = (R \amalg A)/gr(-\alpha)$; then we have $A \subseteq B$, and α can be enlarged to a map $\beta{:}R \to B$. The new module B needn't be injective. It is merely big enough to allow the one map α to be enlarged to a map with domain R. Other maps $\alpha'{:}L' \to A$ may not admit such enlargements. Moreover, for B to be injective we must be able to enlarge maps $\alpha''{:}L'' \to B$. Nevertheless, it is possible to build an injective module containing A by shrewdly repeating the simple construction we have made. If we attempt to make our construction in this manner, we meet the need for infinite ordinals immediately. The approach we will take in the next section will allow us to put off working with infinite ordinals until we absolutely must.

§2 FOR NOETHERIAN RINGS

First, we will expand Observation 1 of §1. Let A be a left R–module and let H be the set of all maps $\alpha{:}L \to A$ for the various left ideals L of R. Let F be the coproduct of copies of R, one for each α in H. We will write $F = \oplus\{R_\alpha \mid \alpha \in H\}$, where each R_α is a submodule of F isomorphic to R. The domain of each map α will be regarded as a submodule of R_α. Then, for each α in H, $gr(-\alpha)$ is a submodule of

R_α II A, which is a submodule of F II A. Let G be the submodule of F II A generated by $\cup\{gr(-\alpha) \mid \alpha \in H\}$.

OBSERVATION 1. The function $\lambda:A \to (F \text{ II } A)/G$ defined by $a\lambda = (0,a) + G$ is an R–monomorphism. For each α in H the function $\beta_\alpha:R_\alpha \text{ II } A \to (F \text{ II } A)/G$ defined by $(r,a)\beta_\alpha = (r,a) + G$ is an R–homomorphism with kernel $gr(-\alpha)$. For each α in H, $\alpha\lambda$ can be enlarged to a map from R to (F II A)/G.

VERIFICATION: We will only comment on the third assertion, since the first two follow from the definitions of G and $gr(-\alpha)$. For a fixed element α of H, we can construct, by the method described in §1, a left R–module B containing A which is isomorphic to $im(\beta_\alpha)$. By Observation 1 of §1, α can be enlarged to a map δ' from R to B and therefore $\alpha\lambda$ can be enlarged to a map $\delta:R \to (F \text{ II } A)/G$. $\square$

Referring to the above observation, we may identify A with its image in $C = (F \text{ II } A)/G$; then we have $A \subseteq C$, and each map $\alpha:L \to A$ can be enlarged to a map $\beta:R \to C$. The new module C needn't be injective. There may be a map $\alpha':L \to C$ that cannot be enlarged to a map with domain R. The present procedure is easy to iterate, and its iteration will yield injective modules.

To describe the iteration procedure, we will start over with fresh notation. Let A_0 be an arbitrary left R–module. Let A_1 be an R–module containing A_0 with the property that every map from a left ideal L of R to A_0 can be enlarged to a map from R to A_1. Let A_2 be a left R–module containing A_1 with the property that each map from a left ideal L of R to A_1 can be enlarged to a map from R to A_2. Continue in this manner to produce a nest of left R–modules $A_0 \subseteq A_1 \subseteq \dots$. Now define $A = \cup A_i$. Is A necessarily injective? It needn't be for arbitrary rings R. However, for one very important class of rings, such an A must be injective.

DEFINITION. A ring R is *left noetherian* if each left ideal of R is finitely generated.

Principal ideal domains are examples of left noetherian rings. If R is a left noetherian ring then so is the ring $R[X]$ of polynomials in an indeterminate X over R. (See page 70 of [Lambek, 1966] for a proof of this.) Thus, for each division ring D, the ring of polynomials in finitely–many indeterminates over D is noetherian. The special case of this result for D a field is known as the Hilbert Basis Theorem, and is a fundamental result in algebraic geometry.

OBSERVATION 2. If R is a left noetherian ring then every left R–module may be embedded in an injective left R–module.

VERIFICATION: We have embedded (above) the arbitrary left R–module A_0 in the union $A = \cup\{A_i \mid i \geqq 0\}$ of left R–modules. We need only verify that A is injective. Let L be an arbitrary left ideal of R and let $\alpha{:}L \to A$ be a map from L to A. Let $\{x_1, ..., x_n\}$ be a finite subset of L which generates L. Each $x_i\alpha$ is contained in one of the modules $A_{k(i)}$. Let $m = \max\{k(1), ..., k(n)\}$. Then $\text{im}(\alpha) \subseteq A_m$. The construction of A_{m+1} provided for an enlargement $\beta{:}R \to A_{m+1} \subseteq A$ of α. Thus, every map $\alpha{:}L \to A$ can be enlarged to a map $\beta{:}R \to A$, and we conclude by Baer's criterion that A is injective. $\square$

EXERCISES

1. Show that a ring R is left noetherian if and only if it contains no countably–infinite nest of distinct left ideals of the form $L_1 \subset L_2 \subset \ ...\ $.

2. Show that the full ring R of all 2×2 matrices of rational numbers is left noetherian. If S is the subring of R

consisting of all matrices of the form $\begin{bmatrix} a & b \\ 0 & d \end{bmatrix}$ with $a \in \mathbb{Z}$ and $b, d \in \mathbb{Q}$, show that S is not left noetherian.

3. Show that a ring R is left noetherian if and only if it contains no countably–infinite sequence of elements $r_1, r_2, ...$ for which the left ideals Rr_1, $Rr_1 + Rr_2$, ... are all distinct.

4. Let R be a principal ideal domain. Show that the class of injective left R–modules has the following closure properties:
(a) Direct summands of injective left R–modules are injective.
(b) The direct product of any family of injective left R–modules is injective.
(c) Factor modules of injective left R–modules are injective.
(d) The coproduct of any family of injective left R–modules is injective.
(e) If A is a left R–module that is the union of a nest of injective submodules then it is injective.

§3 FOR ARBITRARY RINGS

What can be done when the ring R is not left noetherian? In this case we continue to build the nest $A_0 \subseteq A_1 \subseteq ...$ transfinitely. For A_ω we choose $\cup\{A_i \mid i$ a natural number$\}$. For $A_{\omega+1}$ we choose a left R–module containing A_ω and having the property that every map from a left ideal L of R to A_ω can be enlarged to a map from R to $A_{\omega+1}$. In general, the definition is as follows: for each limit ordinal h we define $A_h = \cup\{A_i \mid i < h\}$. For each nonlimit ordinal $h+1$ we choose A_{h+1} to be a left R–module containing A_h and having the property that every map from a left ideal L of R to A_h can be enlarged to a map from R to A_{h+1}. We now have a nest of modules $\{A_h \mid h$ an ordinal$\}$. We will show that there is an ordinal k for which A_k is injective. Our problem now is to think of an ordinal k for which we can imagine a proof that the corresponding module is injective. If we reexamine the left noetherian case, we will see how to choose such a k.

In choosing our k, we will not attempt to make a minimal choice. If m is the cardinality of the ring R, let k be the least ordinal of cardinality greater than m. Then k is a limit ordinal and $A_k = \cup\{A_i \mid i < k\}$. Moreover, for each i < k, the cardinality of i cannot exceed m. We can show that A_k is injective in much the same way that we verified that $A = A_\omega$ is injective when R is left noetherian. Let L be any left ideal of R and let $\alpha:L \to A_k$ be a map. For each x in L, there is an ordinal t(x) < k such that $x\alpha \in A_{t(x)}$. Let $t = \sup\{t(x) \mid x \in L\}$. Since the cardinality of each ordinal t(x) does not exceed m and since the cardinality of L does not exceed m, it follows that the cardinality of the ordinal t does not exceed $m^2 = m$. Consequently, t < k and, furthermore, t+1 < k. We have $im(\alpha) \subseteq \cup\{A_{t(x)} \mid x \in L\} = A_t \subseteq A_{t+1} \subseteq A_k$. By the choice of A_{t+1}, the map α can be enlarged to a map from R to $A_{t+1} \subseteq A_k$. By Baer's criterion, A_k is injective. We have now demonstrated the following observation.

OBSERVATION 1. For any ring R, any left R–module can be embedded in an injective left R–module.

As an example of the value of this observation, we will use it to prove the dual of the corollary to Observation 3 of §6.2. This result will prove to be a convenient tool in §8.3.

OBSERVATION 2. A left R–module P is projective if and only if, for each injective left R–module A and each submodule A' of A, every map $\alpha:P \to A/A'$ can be lifted through the natural map of A onto A/A'.

VERIFICATION: If P is projective this condition is certainly true. Conversely, assume that it is true. Let B be an arbitrary left R–module, B' any submodule of B, and let α be a map from P to B/B'. Embed B as a submodule in an injective left R-module A. Let α' be the composition of α with the embedding B/B' → A/B'. By hypothesis, there is then a map β:P → A such that $\alpha' = \beta\nu$, where ν is the natural map A → A/A'. By the choice of α', we see that im(β) ⊆ B: if a ∈ im(β) and if b is an element of B such that aν = b + B', then a − b ∈ B' ⊆ B and so a = (a − b) + b ∈ B. Therefore β can be considered as a map from P to B. Moreover, if ν':B → B/B' is the natural map then $\beta\nu' = \alpha$ by construction. □

§4 INJECTIVE HULLS

We have seen how an arbitrary left R-module A can be embedded in an injective left R-module C. We made little effort toward economy in this embedding process; that is, the C produced by our construction process will often be far larger than necessary, even for some modules over noetherian rings. We will show that there is a most economical choice for the injective C containing A and that this choice is, in a fairly strong sense, unique. This economic choice will be described by means of the following concept, which will be a valuable tool throughout our study of injective modules.

DEFINITION. A submodule A of a left R-module B is *large* in B if A ∩ A' ≠ (0) for any nonzero submodule A' of B. If A is a large submodule of B we say that B is an *essential extension* of A.

There is standard way of building large submodules of a given module, which we will use often. If B is a submodule of a left R-module A then there always exists a submodule B' of A maximal among all those submodules of A having intersection (0) with B.

OBSERVATION 1. If B is a submodule of a left R–module A and if B' is a submodule of A maximal among those submodules of A having intersection with B equal to (0) then:
(1) B ⊕ B' is a large submodule of A; and
(2) (B ⊕ B')/B' is a large submodule of A/B'.

OBSERVATION 2. If A is a large submodule of a left R–module B and B is a large submodule of a left R–module C, then A is a large submodule of C.

If C is a left R–module and if A ⊆ B and A' ⊆ B' are submodules of C satisfying the conditions that A is large in B and A' is large in B', it does not follow that A + A' is necessarily large in B + B'. To see this, consider the case of R = $\mathbb{Z}$, C = $\mathbb{Z}$ II $\mathbb{Z}$/(2), A = A' = R(2,0), B = R(1,0), and B' = R(1,1). Then A is large in B, and A' is large in B'. On the other hand, A = A + A' is not large in C = B + B' since A ∩ R(0,1) = (0).

Using the notion of a large submodule, we can weaken Baer's criterion.

OBSERVATION 3. A left R–module C is injective if and only if, for each large left ideal L of R and each map α:L → C there is a map β:R → C which restricts on L to α.

VERIFICATION: If C is injective the stated condition is certainly satisfied. Conversely, assume that this condition is satisfied. Let L be a left ideal of R and let α:L → C be a map. Then α has a maximal enlargement α':L' → C for some left ideal L' of R containing L. If L' = R we are done. If not then, by assumption, it is not large in R so there exists a nonzero left ideal I of R such that I ⊕ L' ⊆ R. Now define a map α":I ⊕ L' → C by setting (a + b)α" = bα' for all a in I and b in L'. This map clearly enlarges α, contradicting the maximality of L'. Thus we must have L' = R, and so C is injective by Baer's criterion. □

DEFINITION. An injective left R–module C containing a left R–module A is an *injective hull* of A if A is large in C.

At this stage, of course, we do not know that injective hulls of modules ever exist. However, the following observation will show us where to look to find an injective hull of a left R–module A.

OBSERVATION 4. If a left R–module A is contained in an injective hull C then, for any injective left R–module C' containing A, the inclusion map A → C' can be enlarged to an R–monomorphism from C to C'.

VERIFICATION: By the injectivity of C', the inclusion map A → C' can be enlarged to a map β from C to C'. Since A is large in C and since $\ker(\beta) \cap A = (0)$, we must have $\ker(\beta) = (0)$ and so β is an R–monomorphism. □

This observation tells us that, to find an injective hull for a left R–module A, we may embed A in any way we wish (say as in the preceding sections of this chapter) in an injective left R–module C and then search among the submodules of C containing A. According to Observation 4, if A has an injective hull, at least one such intermediate submodule C' must also be an injective hull of A. What further properties must such an intermediate submodule have if it is to be an injective hull of A? It must contain A as a large submodule. Since it must also be injective, no submodule of C can contain C' as a large submodule. When we put these two requirements together, we have the following observation.

OBSERVATION 5. For a submodule C' of C containing A to be an injective hull of A, it is necessary that C' be maximal among those submodules of C that contain A as a large submodule.

Must there always be at least one submodule of C that satisfies this condition?

OBSERVATION 6. If A is a submodule of a left R-module B then there exists a submodule B' of B that is maximal among all those submodules of B that contain A as a large submodule.

VERIFICATION: Let $\mathcal{F}$ be the family of all submodules of B that contain A as a large submodule. This family is nonempty since A itself certainly belongs to it. Let $\mathcal{M}$ be a maximal nest in $\mathcal{F}$ and let B' be the union of $\mathcal{M}$. Then A is large in B' and, by the maximality of $\mathcal{M}$, no submodule of B that properly contains B' can contain A as a large submodule. □

In our search for an injective hull of an arbitrary left R-module A, our attention has now been narrowed down sufficiently. We are ready to prove that, for any ring R, any left R-module A can be embedded in an injective left R-module C which is an injective hull of A.

OBSERVATION 7. Let C be an injective left R-module containing a left R-module A and let C' be any submodule of C that is maximal among the submodules of C that contain A as a large submodule. Then C' is an injective hull of A.

VERIFICATION: By Observation 4 of §6.2 we need only show that C' is a direct summand of C. Let M be a submodule of C maximal among those submodules the intersection of which with C' is (0). By Observation 1, we note that C' ⊕ M is a large submodule of C and that (C' ⊕ M)/M is a large submodule of C/M. We will show that C =

C' ⊕ M. Let $\nu{:}C \to C/M$ be the natural map and let $\nu'{:}C' \to$ (C' ⊕ M)/M be its restriction to C' ⊕ M. Since C' ∩ M = (0), we see that ν' is in fact an isomorphism and so has an inverse $\mu{:}(C' \oplus M)/M \to C'$. If $\lambda{:}C' \to C$ is the inclusion map then, by the injectivity of C, there exists a map $\beta{:}C/M \to C$ the restriction of which to (C' ⊕ M)/M is just $\mu\lambda$. Now ker(β) ∩ (C' ⊕ M)/M ⊆ ker($\mu\lambda$) = (0) and so ker(β) = (0) since (C' ⊕ M)/M is large in C/M, C' = im(μ) is large in im(β). Since A is large in C' we see, by Observation 2, that A is large in im(β). But C' is maximal among those submodules of C containing A as a large submodule and so we conclude that im(β) = C'. Since β is monic and has the same image as μ, we conclude that (C' ⊕ M)/M = C/M and C' ⊕ M = C, as desired. □

The degree of uniqueness of the injective hull of a module was already implicit in Observation 2 and is spelled out in the following observation.

OBSERVATION 8. If C and C' are injective hulls of a left R–module A then the identity map on A can be enlarged to an isomorphism between C and C'.

VERIFICATION: By the injectivity of C, the inclusion map A → C can be enlarged to a map $\beta{:}C' \to C$. Since A is large in C' and since ker(β) ∩ A = (0), we see that β is a monomorphism. Since im(β) ≅ C' is an injective submodule of C that contains A, we have C = im(β) ⊕ B, from which we conclude that B = (0) since A is large in C. □

EXERCISES

1. Verify Observation 1.

2. Verify Observation 2.

3.   Show that the following conditions on a ring   R   are
equivalent:
(a) A left R-module   C   is injective if and only if every map
from a maximal left ideal of   R   to   C   has an enlargement to a
map from   R   to   C.
(b) Each proper large left ideal   L   of   R   is contained in a left
ideal   H   of   R   such that   H/L   is a simple left R-module.

4.   Let $\alpha:B \to C$ be a homomorphism of left R-modules and
let   A   be a large submodule of   B.
(a) Show that it is not possible in general to conclude that $A\alpha$
is large in $B\alpha$.
(b)   Show   that   if   the   restriction   of α to   A   is   a
monomorphism then α is a monomorphism.
(c) Show that if α is a monomorphism then $A\alpha$ is large in
$B\alpha$.

5.   Let   U   be an independent subset of a left R-module   B,
and let   A   be the submodule of   B   generated by   U. Show
that   A   is a large submodule of   B   if and only if   U   is a
maximal independent subset of   B.

6.   Let $\{B_i \mid i \in \Omega\}$ be a family of left R-modules and, for
each i in Ω,   let A_i be a large submodule of B_i. Show that
$\amalg\{A_i \mid i \in \Omega\}$ is a large submodule of $\amalg\{B_i \mid i \in \Omega\}$.

7.   Let   A   be a left R-module having the property that every
proper submodule of   A   is contained in a maximal (proper)
submodule of   A   which is not large in   A. Show that   A   is
semisimple.

8.   Show that a left R-module is uniform whenever it has a
large uniform submodule.

9.   Let   A   be a left R-module containing submodules   B   and
C   which satisfy $B \cap C = (0)$. Show that if   B   is large in a
submodule   B'   of   A   and if   C   is large in a submodule   C'   of

of A then $B' \cap C' = (0)$.

10. For any left R–module A, let $H_A = \{r \in R \mid rb = 0$ for all elements b of some large submodule B of A$\}$. Show that the following statements are true.
(a) H_A is a two–sided ideal of R.
(b) If B is a submodule of A then $H_B \supseteq H_A$.
(c) If B is a large submodule of A then $H_A = H_B$.
The ideal H_A is called the *tertiary radical* of the module A.

11. Find an infinite set $\{A_i \mid i \in \Omega\}$ of nonzero left R–modules such that $\amalg A_i$ is not a large submodule of ΠA_i.

12. Let D be a division ring and let n be a positive integer. Let S be the full ring of all n$\times$n matrices with entries in D and let R be the subring of S consisting of all upper–triangular matrices (i.e. those matrices having all entries below the main diagonal equal to 0). Then S has the structure of a left R–module.
(a) Let H be the set of all matrices in S having all rows zero except possibly the first row. Show that H is a large R–submodule of S, and is in fact the smallest R–submodule of S.
(b) Use (a) to show that R is a large R–submodule of S.
(c) Use Baer's criterion to show that S is an injective left R–module.
(d) Conclude that S is the injective hull of R as a left module over itself.

§5 DIVISIBLE MODULES

For certain rings it seems natural to approach injectivity from a point of view that is more element–oriented as opposed to the general map-oriented viewpoint. This approach will lead us to the attempt to divide elements of an R-module by elements of R. An element a of a left R-module A is *divisible* by an element r of R if there is an element a' of A for which a = ra'. There is a very blunt limitation on divisibility that is most conveniently described in terms of the following concept, widely used in ring theory.

DEFINITION. The *annihilator* Ann(a) of an element a of a left R-module A is the subset $\{r \in R \mid ra = 0\}$ of R.

OBSERVATION 1. If an element a of a left R-module A is divisible by an element r of R then Ann(r) $\subseteq$ Ann(a).

Because of this observation, we will never consider the possibility of dividing an element a of a left R-module A by an element r of R unless Ann(r) $\subseteq$ Ann(a). The following definition is made with this in mind.

DEFINITION. A left R-module A is *divisible* if for each r of R and each a in A that satisfy Ann(r) $\subseteq$ Ann(a) there is an element a' in A for which a = ra'.

We begin to see the relationship between divisibility and injectivity with the following observation.

OBSERVATION 2. Every injective left R-module is divisible.

VERIFICATION: Let A be an injective left R-module and let r and a be elements of R and A respectively that satisfy Ann(r) $\subseteq$ Ann(a). Define a function α:Rr $\to$ A by (r'r)α = r'a. This is well-defined for if r'r = r"r then (r' − r")r = 0 so r'−r" $\in$ Ann(r) $\subseteq$ Ann(a). Thus (r' − r")a = 0 and so r'a = r"a. It is easy to see that α is a map. By the injectivity of A, it can be enlarged to a map β:R $\to$ A. If a' = 1β then ra' = r(1β) = rβ = rα = a, confirming the divisibility of A. □

For any arbitrary ring R, the divisibility of a module need not imply injectivity, as can be seen from Exercise 8. There is, however, a significant class of rings for which divisibility is equivalent to injectivity.

OBSERVATION 3. If R is a ring having the property that every left ideal is cyclic, then a left R–module A is injective if and only if it is divisible.

VERIFICATION: If A is injective then we have already seen that it must be divisible. Now assume that A is a divisible left R–module. Let L be an arbitrary left ideal of R and let α be a map from L to A. Let r be an element of R satisfying L = Rr. Since Ann(r) $\subseteq$ Ann(rα), there is, by the divisibility of A, an element a' of A satisfying rα = ra'. The map β:R $\rightarrow$ A defined by r'β = r'a' then enlarges α since (r'r)β = r'ra' = r'(rα) = (r'r)α for every r' in R. □

Observation 3 is another one of the key results in this book in which the Axiom of Choice is fundamental. In fact, Blass [1979] has shown that the statement "every divisible **Z**–module is injective" is equivalent, under the Zermelo–Fraenkel axioms of set theory, to the Axiom of Choice.

Some common examples of rings for which every left ideal is cyclic are semisimple rings, principal ideal domains, and proper quotient rings of principal ideal domains (such as **Z**/(n) for any positive integer n).

We will restrict our attention for the duration of this section to modules over a principal ideal domain R. Since R has no proper zero divisors, the restriction placed on division by the annihilator inclusion condition is vacuous; it merely prevents division of a nonzero module element by the zero element of R. In the study of the structure of injective modules in Chapter 9, one of the critical topics will be the determination of injective hulls of certain cyclic modules. In the present circumstances we have a rather convenient way of describing some of these injective hulls. First note that, by Observation 7 of §4, we see that a uniform divisible R-module must be an injective hull of each of its nonzero submodules. Let Q be the field of fractions of the principal ideal domain R. Then Q is certainly a divisible R-module, and it is also uniform since for any two nonzero elements $x = r_1/r_2$ and $y = r_3/r_4$ of Q the cyclic modules Rx and Ry both contain the nonzero element r_1r_3. Thus Q is an injective hull of each of its nonzero submodules. In particular, Q is an injective hull of R. As a particular case of this we note that the ring $\mathbb{Q}$ of rational numbers is an injective hull of the $\mathbb{Z}$-module $\mathbb{Z}$.

For each prime element p of a principal ideal domain R we can construct an additional uniform injective R-module. (In Chapter 9 we will find that we have here constructed, up to isomorphism, every uniform injective R-module.) Notice that for each positive integer n, the cyclic left R-module R/Rp^{n+1} contains a cyclic submodule isomorphic to R/Rp^n. Specifically, $p + Rp^{n+1}$ generates such a submodule. This observation allows us to construct a nest of R-modules as follows. Let $A_1 = R/Rp$. There exists an R-module A_2 isomorphic to R/Rp^2 containing A_1. Similarly, there exists an R-module A_3 isomorphic to R/Rp^3 containing A_2. By continuing in this manner, we produce a countable nest $A_1 \subset A_2 \subset ...$ of R-modules with $A_i \cong R/Rp^i$ for each positive integer i. Let A be the union of this nest of R-modules. We will show that it is divisible and uniform.

Let a be a nonzero element of A and let n be the least positive integer for which a is in A_n. For any nonzero element r of R we have $r = p^k q$, where k is a nonnegative integer and q is an element of R relatively prime to p. By construction, there exists an element y of A_{n+k} satisfying $a = p^k y$. Moreover, if $m = n + k$ then $p^m y = 0$. Since p and q are relatively prime, there exist elements u and v in R satisfying $1 = up^m + vq$. Therefore $y = up^m y + vqy = q(vy)$ and so $a = p^k y = p^k q(vy) = r(vy)$. From this we conclude that A is divisible. The cyclic submodule generated by a is $A_n \supseteq A_i$; consequently, since a is an arbitrary nonzero element of A, we see that A is uniform.

Since A is uniform and divisible (that is, injective), it is an injective hull of each of the cyclic submodules $A_i \cong R/Rp^i$. We will adopt $R(p^\infty)$ as the standard notation for any left R—module isomorphic to A.

We leave the verification of the following observation as an exercise.

OBSERVATION 4. Let p be a prime element of a principal ideal domain R. Then each nonzero finitely—generated submodule of $R(p^\infty)$ is isomorphic to R/Rp^n for some nonnegative integer n.

EXERCISES

1. Show that, for an arbitrary ring R, the class of divisible left R—modules is not closed under taking factor modules.

2. Let $\mathbb{Z}/(n)$ be the ring of integers modulo n, and regard it as a module over itself in the usual way.
(a) If n is a power of a prime, show that $\mathbb{Z}/(n)$ is divisible and therefore injective as a $\mathbb{Z}/(n)$-module.
(b) Show that $\mathbb{Z}/(n)$ is divisible (and therefore injective) even when n is not a power of a prime.

3. Let R be a principal ideal domain and let I be a nonzero ideal of R. Regard the ring R/I as a module over itself in the usual way. Show that it is divisible and therefore injective (as an R/I—module).

4. Let p be a prime element of a principal ideal domain R.
(a) Show that the cyclic submodule of $R(p^\infty)$ generated by an element a of $R(p^\infty)$ is the least member of the nest defining $R(p^\infty)$ that contains a.

(b) Show that the submodule of $R(p^\infty)$ generated by a finite set of elements of $R(p^\infty)$ is cyclic.
(c) Use (a) and (b) to verify Observation 4.

5. Let R be a principal ideal domain and let F be the field of fractions of R. For a prime element p of R, let $F(p) = \{r/p^n \mid r \in R$ and n a nonnegative integer}. Observe that $F(p)$ is an R-submodule of F which contains R. For any nonnegative integer n, the inclusions $R \subseteq F(p) \subseteq F$ yield $R/Rp^n \subseteq F(p)/Rp^n$. Show that $F(p)/Rp^n$ is an injective hull of R/Rp^n if the integer n is positive.

6. Let p be a prime integer. For each positive integer n, let $\mathbb{C}(p^n)$ be the set of all those complex numbers that are p^nth roots of 1.
(a)   Verify that, for each n, $\mathbb{C}(p^n)$ is a subgroup of the multiplicative group of complex numbers having absolute value 1.
(b) Verify that $\mathbb{C}(p^n)$ is cyclic of order p^n and that $\mathbb{C}(p) \subset \mathbb{C}(p^2) \subset \dots$.
(c) Let $\mathbb{C}(p^\infty) = \cup\{\mathbb{C}(p^n) \mid n$ a positive integer}. Show that $\mathbb{C}(p^\infty)$ is an injective hull of $\mathbb{C}(p^n)$ for each n. (Here we are regarding $\mathbb{C}(p^\infty)$ and its subgroups as $\mathbb{Z}$-modules, though the notation is multiplicative rather than additive.)

7. If U is a nonempty subset of a left R-module A then we define Ann(U) to be $\cap\{Ann(a) \mid a \in U\}$. Suppose that R is a ring which is injective as a left module over itself and suppose that I and H are left ideals of R. Show that Ann(I $\cap$ H) = Ann(I) + Ann(H).

8. Let $R = \mathbb{Z}[X]$ be the ring of polynomials over $\mathbb{Z}$ and let K be its field of quotients. Show that K/R is a divisible left R-module which is not injective.

NOTES FOR CHAPTER 7

The ability of an arbitrary module to be embedded in an injective module is a major feature of module theory. This result, due to Baer [1940], was first exploited systematically by Cartan and Eilenberg [1956]. The equally–fundamental theory of the injective hull is due to Eckmann and Schopf [1953]. A slicker (but less intuitive) proof of the embeddability of every module in an injective module is given in Rotman [1970]. S. Wiegand [1972] has shown an interesting analogy between algebraic extensions of fields and essential extensions of modules, in which the role of the algebraic closure of a field is played by the injective hull of the module. She has used this analogy to create the rudiments of a "Galois theory" of essential extensions of modules.

The dual notion of the injective hull is that of the projective cover of a module. See Bass [1960]. Unfortunately, projective covers of modules do not always exist for arbitrary rings. Rings over which all modules have projective covers are considered so wonderful by algebraists that they are called "perfect rings".

The notions of injective objects and injective hulls can be defined in contexts (categories) other than those discussed here, and are equally important there. In the category of normal topological spaces, Tietze's Theorem states that the closed interval [0,1] is injective; see Kelley [1955]. In the category of Banach spaces, the fact that the space of real numbers is injective is known as the Hahn–Banach Theorem; refer to Rudin [1966].

8

Injective Modules as a Class

§1 GENERAL CLOSURE PROPERTIES

Before we begin our structural investigation of injective modules, we will determine the closure properties of the class of injective R–modules for various rings R. For an arbitrary ring R the class of injective modules has only two closure properties of significance. One of them we have already seen in Observation 4 of §6.2: the class of injective modules is closed under taking direct summands. The second is given by the following observation.

OBSERVATION 1. The direct product $\Pi\{C_i \mid i \in \Omega\}$ of a family $\{C_i \mid i \in \Omega\}$ of left R–modules is injective if and only if each C_i is injective.

VERIFICATION: Suppose that each left R–module C_i is injective. Let A be a submodule of a left R–module B and let $\alpha{:}A \to \Pi C_i$ be a map. For each h in Ω, let $\pi_h{:}\Pi C_i \to C_h$ be the canonical projection map onto C_h. Then, by the injectivity of C_h, we know that there exists a map $\beta_h{:}B \to C_h$

the restriction of which to A is $\alpha \pi_h$. Therefore there exists a map $\beta : B \to \Pi C_i$ such that $\beta \pi_h = \beta_h$ for each h in Ω and the restriction of β to A is precisely α. $\square$

Notice an important consequence: the coproduct of a *finite* family of injective modules is always injective.

§2   COGENERATORS

We have the proper tools to construct, for each ring R, modules that have properties that are dual to those of generators. Recall that in Chapter 6 we defined an R–module G to be a generator if and only if every R–module is an epimorphic image of a coproduct of copies of G.

DEFINITION. A left R–module C is a *cogenerator* if every left R–module is isomorphic to a submodule of a direct product of copies of C.

One consequence of this definition is immediate. A left R–module containing a cogenerator as a submodule is itself a cogenerator. As for generators, one always has the simple example of a generator that is also projective: R. In a few special cases, a cogenerator is also readily available. For example, if R is semisimple then R is a cogenerator that is also injective. For any arbitrary ring, it is not as evident that a cogenerator exists. We will show this, and also show that there exists a cogenerator which is "minimal" in a very well–defined sense.

OBSERVATION 1. A left R–module C is a cogenerator if and only if, for each simple left R–module A, C contains a submodule isomorphic to an injective hull of A.

VERIFICATION: First assume that C is a cogenerator. Let A be a simple left R–module having an injective hull E. Since C is a cogenerator, E is isomorphic to a submodule E' of a direct product of copies of C. Furthermore, E' in turn has a large submodule A' isomorphic to A. As a consequence of Observation 1 of §0.6, there must be a projection π of E onto one of the factors in this direct product the kernel of which does not contain A'. Since A' is simple, this means that $\ker(\pi) \cap A' = (0)$ and since A' is large in E', we can conclude that $\ker(\pi) = (0)$. Thus C contains the submodule $\text{im}(\pi)$ isomorphic to an injective hull of A.

Conversely, suppose that C is a left R–module satisfying the stated condition and let A be an arbitrary nonzero left R–module. For each nonzero element a of A we will construct a map α_a from A to C satisfying $a\alpha_a \neq 0$. To do this is not hard: given the element a, let B_a be the union of a maximal nest of submodules of A each of which does not contain a. This is a submodule of A to which a does not belong. Moreover, $B = [B_a + Ra]/B_a$ is a simple submodule of A/B_a. By hypothesis, C contains a submodule E isomorphic to an injective hull of B and so there exists an R–monomorphism from B to E which, by injectivity, can be extended to a map β_a from A/B_a to E. Now define α_a to be composition of the natural map $A \to A/B_a$, the map β_a, and the inclusion of E into C. Then surely $a\alpha_a \neq 0$.

For each nonzero element a of A let C_a be a copy of C. Define a map α from A to $\Pi\{C_a \mid 0 \neq a \in A\}$ as follows: if $b \in A$ and if $0 \neq a \in A$ then $(b\alpha)(a) = b\alpha_a$. The kernel of this map is (0) since, for each nonzero element a of A, we have $(a\alpha)(a) \neq 0$. Thus we have shown that every left R–module can be embedded in a product of copies of C, proving that C is a cogenerator. $\square$

Let $\{A_i \mid i \in \Omega\}$ be a set of nonisomorphic simple left R–modules chosen such that every simple left R–module is isomorphic to precisely one of them. For each i in Ω, let D_i be an injective hull of A_i. Finally, set $D = \amalg\{D_i \mid i \in \Omega\}$. By

Observation 1, D is a cogenerator since an injective hull of any simple left R–module is isomorphic to some submodule of D. In fact, D is a minimal cogenerator in the sense that every cogenerator must contain a submodule that is isomorphic to D. This will be easily demonstrated after noting the following:

OBSERVATION 2. Let $B_1 \oplus \ldots \oplus B_k$ be a direct sum of submodules of a left R–module B and, for each $1 \leq i \leq k$, let A_i be a submodule of B containing B_i as a large submodule. Then the sum ΣA_i of submodules of B is also direct.

We now show that if C is a cogenerator then D is isomorphic to a submodule of C. Indeed, for each i in Ω there is a set Γ and embedding β of D_i into the direct product $\{C_h \mid h \in \Gamma\}$ of copies of C. In particular, for some element h of Γ the kernel of the projection π_h of this direct product onto the factor C_h does not contain the image of A_i under β, for otherwise A_i would be in the kernel of β, which it is not. Thus $\ker(\beta\pi_h) \cap A_i$ is a proper submodule of A_i and so must equal (0) by the simplicity of A_i. Since A_i is large in D_i, this means that $\ker(\beta\pi_h) = (0)$ and so D_i is isomorphic to a submodule C_i of C. Let B_i be the submodule of C_i that is isomorphic to A_i. We are done if we can show that the sum $\Sigma\{C_i \mid i \in \Omega\}$ is direct, for then it is surely a submodule of C isomorphic to D. Assume that this is not the case; then $\Sigma\{C_j \mid j \in \Lambda\}$ is not direct for some finite subset Λ of Ω, which, by Observation 2, contradicts the fact that the sum $\Sigma\{B_i \mid i \in \Lambda\}$ is direct by Observation 1 of §2.3.

We now mention a very subtle point, so subtle that it was missed in many standard textbooks on module theory. While it is true that D is a cogenerator having the property that every cogenerator must contain a submodule that is isomorphic to D, there may well exist other cogenerators having this same property *which are nonetheless not isomorphic to* D. The first examples of rings over which this occurs were given by Osofsky [1990].

We also note that, as a consequence of the above, there is an injective cogenerator which is isomorphic to a submodule of any other injective cogenerator, namely an injective hull of D. For some important classes of rings the module D is itself injective. For example, from Observation 1 in §4 it will follow that this is true for left noetherian rings.

EXERCISES

1. Use a finite induction argument to verify Observation 2.

2. Show that the $\mathbb{Z}$-module $\mathbb{Q}/\mathbb{Z}$ is a cogenerator by showing that it contains submodules isomorphic to all simple abelian groups.

3. Generalize Exercise 2 by showing that if R is a principal ideal domain having a field of fractions $F \neq R$ then F/R is a cogenerator.

4. A left R-module A is *cofinitely generated* if and only if it has an injective hull of the form $C_1 \amalg \ldots \amalg C_n$, where the C_i are injective hulls of simple left R-modules.
(a) Let A be a left R-module and let A' be the sum of all simple submodules of A. Show that A is cofinitely generated if and only if A' is a large finitely-generated submodule of A.
(b) Show that a left R-module A is cofinitely generated if and only if, for any embedding of A into a direct product $\Pi\{B_i \mid i \in \Omega\}$ of left R-modules there exists a finite subset Λ of Ω such that the image of A is in $\Pi\{B_i \mid i \in \Lambda\}$.
(c) Show that the left $\mathbb{Z}$-module $\mathbb{Z}$ is finitely generated but not cofinitely generated.
(d) If $A = A_1 \amalg \ldots \amalg A_n$, show that A is cofinitely generated if and only if each of the A_i is cofinitely generated.

5. Let $\varphi{:}R \to S$ be a homomorphism of rings, and let S have

the structure of a left R–module as in Exercise 5 of §6.2. Then $\text{Hom}_R(S,A)$ has the structure of a left S–module for every left R–module A. Show that this is a cogenerator if A is.

§3   HEREDITARY RINGS AGAIN

When we ask for those rings for which every factor module of an injective left R–module is injective, we are led back to a familiar class.

OBSERVATION 1. A ring R has the property that every factor module of an injective left R–module is injective if and only if it is left hereditary.

VERIFICATION: Suppose that R is left hereditary and let A' be a submodule of an injective left R–module A. We must show that A/A' is also injective. Let L be a left ideal of R and let α be a map from L to A/A'. Since R is left hereditary, L is a projective left R–module and so there exists a map $\beta{:}L \to A$ such that $\alpha = \beta\nu$, where $\nu{:}A \to A/A'$ is the natural map. Since A is injective, there is a map β' from R to A the restriction of which to L is β. Then the restriction of $\beta\nu$ to L is α, proving that A/A' is injective.

Now suppose that R has the property that factor modules of injective left R–modules are always injective. Let A be an arbitrary injective left R–module, let A' be a submodule of A, let L be a left ideal of R, and let α be a map from L to A/A'. By Observation 2 of §7.3, we see that to prove that L is projective we need only show that α can be lifted through the natural map ν of A onto A/A'. Since A/A' is injective, we know that α has an enlargement $\beta{:}R \to A/A'$. Since R is projective, there exists a map $\beta'{:}R \to A$

satisfying $\beta = \beta'\nu$. If $\lambda:L \to R$ is the inclusion map, then $\lambda\beta'$ is a map from L to R satisfying $(\lambda\beta')\nu = \alpha$. $\square$

For which rings R is every left R–module isomorphic to a factor module of an injective left R–module? The answer to this question is the class of quasi–Frobenius rings, which will be discussed briefly in §6 of Chapter 9.

EXERCISES

1. Let R be a left hereditary ring and let $C_1, ..., C_n$ be injective submodules of a left R–module A. Show that ΣC_i is an injective submodule of A.

2. Let R be a ring contained in a left R–module C. Let A be an arbitrary left R–module.
(a) Show that A is isomorphic to a submodule of $\amalg\{C_i \mid i \in \Omega\}/B$, for some family $\{C_i \mid i \in \Omega\}$ of copies of C and some submodule B of this coproduct.
(b) Suppose further that R is left hereditary and that C is injective. Show that A can be embedded in an injective module of the form $\Pi\{C_i \mid i \in \Omega\}/B$.

3. Show that a ring R is left hereditary if and only if, for any left R–module A, the sum of two injective submodules of A is again injective.

§4 NOETHERIAN RINGS

For which rings is the class of injective left R–modules closed under taking coproducts? Not all rings. For example, let F be a field and let R be a direct product of countably–many copies of F. For each positive integer i, let r_i be the element of R defined by $r_i(j) = 0$ if j does not equal i and $r_i(j) = 1$ if $j = i$. Then Rr_i is an injective left R–module for each i, but $\amalg\{R_i \mid i$ a positive integer$\}$ is not injective. The answer to our question will take us back to the class of left noetherian rings introduced in §7.2.

OBSERVATION 1. If R is a left noetherian ring and if $\{C_i \mid i \in \Omega\}$ is a family of injective left R–modules, then $\amalg\{C_i \mid i \in \Omega\}$ is injective.

VERIFICATION: Baer's criterion is a powerful tool here. Let L be a left ideal of R and let α be a map from L to $C = \amalg\{C_i \mid i \in \Omega\}$. Since R is left noetherian, L is finitely generated and so, by the nature of the coproduct, there is a finite subset Λ of Ω such that $im(\alpha) \subseteq C' = \amalg\{C_i \mid i \in \Lambda\}$. We know that a coproduct of finitely–many injective left R–modules is injective, and so there exists a map $\beta:R \to C'$ which enlarges α. The composition of this map with the inclusion map $C' \to C$ is then a map which restricts to α on L. $\square$

OBSERVATION 2. If R is a ring having the property that the coproduct of every countable family $\{C_i \mid i$ a positive integer$\}$ of injective hulls of simple left R–modules is injective, then each submodule of a finitely–generated left R–module is finitely generated.

VERIFICATION: Let R have the property mentioned in the observation, and let A be a submodule of a finitely–generated left R–module B. We will verify the observation by assuming that A is not finitely generated and producing a contradiction: since A is not finitely generated, elements $\{a_i \mid i$ a positive integer$\}$ can be chosen in succession from A so that $Ra_1 \subset Ra_1 + Ra_2 \subset ...$ is a strictly ascending nest of submodules of A. For each i, let $A_i = Ra_1 + ... + Ra_i$, and let $A' = \cup A_i$. For each x in A' there is an n such that x is in A_j for all $j \geq n$. Thus we can define a map α from A' to $\amalg\{A'/A_i \mid i$ a positive integer$\}$ by $x\alpha = (x+A_1, x+A_2, \ldots)$. As a consequence of Observation 1 of §2 we note that for each i there is a nonzero map θ_i from A'/A_i to a left R-module C_i which is the injective hull of a simple left R-module. These maps then define a map θ from $\amalg A'/A_i$ to $\amalg C_i$. Thus we have a map $\alpha\theta$ from A' to $\amalg C_i$. By the hypothesis of the observation, $\amalg C_i$ is injective and α may be enlarged to a map β from B to $\amalg C_i$. For each natural number h, let π_h be the projection from $\amalg C_i$ onto C_h. Then we have two contradictory statements concerning $\text{im}(\alpha)$: since B is finitely generated, $B\beta\pi_h = (0)$ for all but finitely-many indices h; on the other hand, for each h we have $B\beta\pi_h \supseteq A'\alpha\theta\pi_h \neq (0)$. This verifies the observation. $\square$

OBSERVATION 3. A ring R has the property that the coproduct of every countable family of injective hulls of simple left R-modules is injective if and only if it is left noetherian.

VERIFICATION: If R is left noetherian then the required closure property follows from the stronger one stated in Observation 1. If R has the property that the coproduct of every countable family of injective hulls of simple left R-modules is injective, then each left ideal of R must be finitely generated by Observation 2. $\square$

In particular, a ring R has the property that the coproduct of every countable family of injective left R-modules is injective if and only if it is left noetherian. We also note that the injective modules over a left noetherian ring have another simple closure property.

OBSERVATION 4. If R is a left noetherian ring and if C is a left R-module which is the union of a nest $\{C_i \mid i$

$\in \Omega\}$ of injective submodules then C is injective.

VERIFICATION: Let α be a map from a left ideal L of R to C. Since L is finitely generated and since $\{C_i \mid i \in \Omega\}$ is a nest, there is an element j of Ω such that $\text{im}(\alpha) \subseteq C_j$. By the injectivity of C_j, the map α can be enlarged to a map $\beta\colon R \to C_j \subseteq C$. $\square$

OBSERVATION 5. A ring R has the property that the union of each countable nest of injective left R-modules is injective if and only if it is noetherian.

VERIFICATION: If R is noetherian, then the required closure property holds as a consequence of Observation 4. Now suppose that R has the stated closure property. If $\{C_i \mid i$ a positive integer$\}$ is a countable family of injective left R-modules then $C_1 \subseteq C_1 \amalg C_2 \subseteq \ldots$ is a countable nest of injective left R-modules the union of which is $\amalg C_i$. By hypothesis, this is injective. By Observation 3, R is left noetherian. $\square$

EXERCISES

1. Let R be a left noetherian ring and let A be a left R-module.
(a) Show that A contains a maximal injective submodule C. Can you determine whether C must be unique?
(b) Show that $A = B \oplus C$, where C is injective and B contains no nonzero injective submodules.
(c) Show that if R is both left noetherian and left hereditary then A contains a unique maximal injective submodule.

2. Let R be an arbitrary ring.
(a) Let A be a left R-module that contains a finite independent set $\{m_1, \ldots, m_k\}$ which generates a large submodule of A. For each $1 \leq i \leq k$, let A_i be an injective hull for

the cyclic submodule Rm_i of A. Show that A is isomorphic to a submodule of $II\{A_i \mid 1 \leq i \leq k\}$.

(b) Now assume that R is left noetherian and that A is a left R-module having a maximal independent subset B (not necessarily finite). For each b in B, let A_b be an injective hull of the cyclic left R-module Rb. Show that A is isomorphic to a submodule of $II\{A_b \mid b \in B\}$.

3. Let R be an arbitrary ring and let $\{C_i \mid i \in \Omega\}$ be a collection of injective left R-modules. Show that $II\{C_i \mid i \in \Omega\}$ is injective if and only if, for each countably-infinite nest $L_1 \subseteq L_2 \subseteq \ldots$ of left ideals of R there exist a positive integer n and a finite subset Λ of Ω such that $Hom_R(L_{j+1}/L_j, E_i) = (0)$ for all $j > n$ and all $i \in \Omega \setminus \Lambda$.

4. Let R be a left hereditary left noetherian ring and let $\{C_i \mid i \in \Omega\}$ be a collection of injective submodules of a left R-module A. Show that ΣC_i is an injective submodule of A.

5. Let R be a left hereditary left noetherian ring and let C be an injective left R-module containing R. Show that any left R-module A can be embedded in an injective module of the form $II\{C_i \mid i \in \Omega\}/B$, for some family $\{C_i \mid i \in \Omega\}$ of copies of C and some submodule B of this coproduct.

6. Let R be a left noetherian ring and let A be a finitely-generated right R-module. Show that $Hom_R(A,R)$ is a noetherian left R-module.

NOTES FOR CHAPTER 8

In summary, we see that the class of injective left R-modules is closed under:

(a)   summands for all R;
(b)   direct products for all R;
(c)   factor modules if and only if R is left hereditary;
(d)   coproducts if and only if R is left noetherian;
(e)   unions of nests if and only if R is left noetherian; and
(f)   submodules if and only if R is semisimple.

The observation that the closure of the class of injectives under coproducts (or unions of nests) necessitates the ring being left noetherian is due to Hyman Bass and Zoltan Papp, working independently. The remaining closure properties can be found in Cartan & Eilenberg [1956].

9

Structure of Injective Modules

§1 INTRODUCTION

We have come to the climax of Part Two. Three tools will be fundamental for our structural investigations: (1) the concept of a uniform module introduced in Chapter 1, (2) our knowledge of injective hulls, and (3) some special features of modules over left noetherian rings. A reasonably good insight into the strucure of uniform injective modules can be developed. For arbitrary injective modules, the structure will only be clarified when the ring is left noetherian.

Twice before we have seen that left noetherian rings behave especially well in relation to injectivity. The construction of injective hulls is appreciably easier when the ring is left noetherian. For left noetherian rings, the class of injective left R—modules is closed under taking coproducts. This closure property and the good behavior to be exhibited in §5 with respect to uniformity are the keys to the determination of the structure of the injective modules over left noetherian rings.

We will give no structural results for arbitrary injective left R—modules over rings which are not left noetherian. The device that we consistently use for analyzing the structure of modules is the method of direct sum decomposition. A few moments of consideration will suggest that a penetrating analysis of the structure of injective modules by this means could be expected only for rings for which the class of injective modules is closed under coproducts. As we have seen, these rings are precisely the left noetherian rings. To determine the structure of injectives over more general rings, other methods will be needed.

§2   UNIFORM INJECTIVE MODULES

It is possible to reduce the problem of determining the structure of uniform injective modules over an arbitrary ring to considerations involving the left ideals of the ring. This possibility is suggested by the principle that each nonzero element of a uniform module is interrelated with the overall structure of the module. In the present situation this principle takes the following form.

> **OBSERVATION** 1. If c is a nonzero element of a nonzero uniform injective left R—module C then Rc is a uniform cyclic left R—module and C is an injective hull of Rc.

> **VERIFICATION**: Since C is uniform, Rc is uniform and large in C. Since C is injective, it must be an injective hull of Rc. □

The observation suggests a program for cataloging the uniform injectives over R. The first step should be to catalog the uniform cyclic left R—modules. Since each cyclic left R—module is isomorphic to a factor R/L for some left ideal L of R, this step merely requires determining which left ideals of R produce uniform factor modules. The second step should be

to determine when two uniform left R–modules R/L and R/L'
have isomorphic injective hulls. We begin by answering the
following question: for which left ideals L of R is R/L
uniform?

DEFINITION. A left ideal L of a ring R is
meet–irreducible if it is not the intersection of any two left
ideals of R properly containing it.

OBSERVATION 2. For a left ideal L of R, the module
R/L is uniform if and only if L is meet–irreducible.

VERIFICATION: If R = L, the observation is immediate.
Therefore, for the remainder of the verification, we will assume
that L is proper. Suppose that R/L is uniform and that L'
and L" are left ideals that properly contain L. Then the
intersection of the nonzero submodules L'/L and L"/L of
R/L is again nonzero. This means that L' ∩ L" ≠ L. Now
suppose, conversely, that L is meet–irreducible. Any pair of
nonzero submodules of R/L are of the form L'/L and L"/L,
where L' and L" are left ideals of R properly containing L.
By the meet–irreducibility of L, L' ∩ L" ≠ L and so (L'/L) ∩
(L"/L) = (L' ∩ L")/L, which is not (0). □

When do uniform cyclic R–modules have isomorphic
injective hulls?

OBSERVATION 3. Let A be a nonzero uniform cyclic
left R–module having an injective hull C and let A' be a
nonzero uniform cyclic left R–module having an injective
hull C'. Then C ≅ C' if and only if there are nonzero
submodules B and B' of A and A' for which B ≅
B'.

VERIFICATION: Suppose that α:C → C' is an
isomorphism. Since C' is uniform, the submodule B' = A' ∩
im(α) is nonzero, and is isomorphic to the submodule B =

B'α^{-1} of A. Conversely, suppose that α:B $\rightarrow$ B' is an isomorphism between submodules B of A and B' of A'. By injectivity, this map has an enlargement β:C $\rightarrow$ C'. Since B is large in A, it is large in C as well. Since ker(β) $\cap$ B = (0), this means that ker(β) = (0) and so β is a monomorphism. Then im(β) is injective and from the uniformity of C' we conclude that β is an isomorphism. $\square$

Uniform modules are always indecomposable. We should notice at this stage that the converse holds if the modules are injective. Indeed, let C be an indecomposable injective left R-module and let B' be a nonzero submodule of C. Then B' has an injective hull C' contained in C, which is a direct summand of C. Since C is indecomposable, this means that C' = C and so B' is large in C. Therefore there can be no nonzero submodule B" of C satisfying B' $\cap$ B" = (0) and so no submodule of C can be a direct sum of proper submodules. This remark allows us to make the following observation.

OBSERVATION 4. A nonzero injective left R-module is uniform if and only if its endomorphism ring is local.

VERIFICATION: By Observation 5 of §1.2 we know that a left R-module having a local endomorphism ring is indecomposable and so one direction of the observation follows from the above remarks. Now assume that A is a uniform injective left R-module. If α is an endomorphism of A having kernel (0) then Aα is an injective large submodule of A and so it must equal all of A. This implies that α is an automorphism of A and so, in particular, is invertible. We conclude therefore that the noninvertible endomorphisms of A are precisely those maps having nonzero kernels. If α and β are maps of this kind then, by uniformity, ker(α) $\cap$ ker(β) $\neq$ (0). But ker(α + β) $\supseteq$ ker(α) $\cap$ ker(β), and so α + β is also noninvertible. This proves that the endomorphism ring of A is local. $\square$

We can now give an explicit description of our cataloging procedure. Let $\mathcal{L}$ be the set of all meet–irreducible left ideals of R. For L and L' in $\mathcal{L}$, we write L $\sim$ L' if R/L and R/L' contain isomorphic nonzero submodules. From Observation 3 we know that $\sim$ is an equivalence relation on $\mathcal{L}$ and that the injective hulls of R/L and R/L' are isomorphic if and only if L $\sim$ L'. Then a complete catalog, without repetitions, of the uniform injective left R–modules over a ring R may be obtained as follows: from each equivalence class of meet–irreducible left ideals of R, choose one member and take its injective hull. This cataloging procedure will be illustrated for a very special type of ring in the next section.

EXERCISES

1. Show that every simple left R–module is injective if and only if every left ideal of R is the intersection of maximal left ideals.

2. If R is a ring satisfying the equivalent conditions in Exercise 1, show that no proper cyclic left ideal of R is isomorphic to R.

3. If R is a commutative ring, show that every simple left R–module is injective if and only if every finitely–generated ideal of R is of the form Re for some idempotent element e of R.

§3 SEMIUNIFORM INJECTIVE MODULES

If A is a semiuniform module we know that it can be written as a direct sum $\oplus\{A_i \mid i \in \Omega\}$ of uniform submodules.

In general, there is no reason for this representation to be unique in any significant sense, but if the module A happens to be injective as well, it turns out that we can say quite a bit about the uniqueness of such a representation.

We begin with a technical observation.

OBSERVATION 1. Let E be an injective left R-module which can be written as $B \oplus B'$ and $\oplus\{A_i \mid i \in \Omega\}$, where B and the A_i are all nonzero uniform submodules of E, and where associated with these direct-sum decompositions we have inclusion maps $\lambda:B \to E$ and $\lambda_i:A_i \to E$ (for each i in Ω) and projections $\pi:E \to B$ and $\pi_i:E \to A_i$ (for each i in Ω). Then there exists an element h of Ω such that the $\lambda_h\pi$ is an isomorphism and $E = A_h \oplus B'$.

VERIFICATION: Since B is a direct summand of E, it is injective. If b is a nonzero element of B there exists a finite subset Λ of Ω such that $b \in D_1 = \oplus\{A_i \mid i \in \Lambda\}$. Then $E = D_1 \oplus D_2$, where $D_2 = \oplus\{A_i \mid i \in \Omega \setminus \Lambda\}$. For $j = 1,2$ let $\eta_j:D_j \to E$ and $\theta_j:E \to D_j$ be the canonical injection and projection, respectively. Then $\lambda\theta_j\eta_j\pi$ is an endomorphism of B for each j and the sum of these two endomorphisms is the identity map on B. Since the endomorphism ring of B is local by Observation 4 of §2, we conclude that one of these maps must be an isomorphism. Indeed, it must be $\lambda\theta_1\eta_1\pi$ since $b\lambda\theta_2\eta_2\pi = 0$. For each i in Λ, let $\beta_i = \lambda\pi_i\lambda_i\pi$. Each of these maps is again an endomorphism of B and the sum of the β_i is $\lambda\theta_1\eta_1\pi$. As before, the locality of the endomorphism ring of B implies that there exists an element h of Λ such that β_h is an isomorphism.

Let $\gamma = \lambda\pi_h$ and $\delta = \lambda_h\pi$. Then $\beta_h = \gamma\delta$ and so, in particular, δ is an epimorphism. If $a \in A_h$ then $a\delta \in B$ so $a\delta = b\gamma\delta$ for some $b \in B$. Therefore $a = b\gamma + (a - b\gamma) \in$

$B\gamma + \ker(\delta)$. Since $B\gamma \cap \ker(\delta) = (0)$, we see that $A_h = B\gamma \oplus \ker(\delta)$. But A_h is uniform and $B\gamma \neq (0)$, so we conclude that $\ker(\delta) = (0)$, which proves that δ is an isomorphism, as desired.

If $b \in B$ then there exists an element a of A_h such that $b\pi = b = a\delta = a\pi$ and so $b - a \in \ker(\pi) = B'$. Thus $b \in A_h + B'$ for all such elements b, proving that $E = A_h + B'$. On the other hand, $A_h \cap B' = \ker(\delta) = (0)$ and so $E = A_h \oplus B'$. $\square$

Before going on, we need to mention two facts about sets. The first of these is rather obvious: if Ω is an infinite set and if, to each element i in Ω, we associate a finite set A_i such that not all of these sets are empty, then there is an injective function from $\cup\{A_i \mid i \in \Omega\}$ to Ω. The second fact is more subtle: if A and B are nonempty sets and if there exist injective functions $A \to B$ and $B \to A$ then there is, in fact, a one-to-one correspondence between A and B. This result is known as the Schroeder–Bernstein Theorem and is proven in most introductory textbooks on set theory. An analog of this theorem for injective modules, due to Bumby [1965], is given in Exercises 1 and 2.

OBSERVATION 2. Let E be a semiuniform injective left R–module and let $\{A_i \mid i \in \Omega\}$ and $\{B_j \mid j \in \Lambda\}$ be families of uniform submodules of E satisfying the condition that $E = \oplus\{A_i \mid i \in \Omega\} = \oplus\{B_j \mid j \in \Lambda\}$. Then there exists a bijective function θ from Ω to Λ satisfying the condition that $A_i \cong B_{\theta(i)}$ for all i in Ω.

VERIFICATION: We begin by defining an equivalence relation $\sim$ on the set Ω by setting $i \sim h$ if and only if the modules A_i and A_h are isomorphic. A similar relation can be defined on Λ. If $i \in \Omega$ then, by Observation 1, we see that there is an element j of Λ such that A_i and B_j are isomorphic. This means that there must exist a one-to-one correspondence between the set of equivalence classes of Ω defined by the relation $\sim$ and the corresponding set of equivalence classes of Λ. We will be done if we can find a bijective function θ between any pair of such corresponding classes.

In other words, we can assume that we are considering nonempty subsets Ω' of Ω and Λ' and Λ satisfying the following conditions:
(1) If $i \in \Omega'$ and $h \in \Omega$ then $A_i \cong A_h$ if and only if $h \in \Omega'$;
(2) If $j \in \Lambda'$ and $k \in \Lambda$ then $B_j \cong B_k$ if and only if $k \in \Lambda'$;
(3) If $i \in \Omega'$ and $j \in \Lambda'$ then $A_i \cong B_j$.

Pick an element $j(1)$ of Λ'. By Observation 1, there exists an element $i(1)$ of Ω' satisfying $E = A_{i(1)} \oplus [\oplus \{B_j \mid j(1) \neq j \in \Lambda\}]$. Again, if $j(2)$ is an element of $\Lambda' \setminus \{j(1)\}$ then there exists an element $i(2)$ of $\Omega' \setminus \{i(2)\}$ satisfying $E = A_{i(1)} \oplus A_{i(2)} \oplus [\oplus \{B_j \mid j(1), j(2) \neq j \in \Lambda\}]$. We can continue in this manner. If the set Λ' is finite then we will obtain a corresponding finite subset of Ω' and so show that Ω' has at least as many elements as Λ'. We can also reverse the roles of Ω' and Λ' in the above argument and so conclude that whenever at least one of the sets Λ' and Ω' is finite then both of them are finite and have the same number of members, so that the desired bijective function exists.

Now assume both of the sets Ω' and Λ' are infinite. Choose an element i of Ω' and, for each element j of Λ, let α_j be the restriction of the projection $E \to B_j$ to A_i. Let Λ_i be the subset of Λ consisting of all those elements j such that α_j is an isomorphism. Clearly $\Lambda_i \subseteq \Lambda'$ for each i. On the other hand, by Observation 1 we see that each element of Λ' belongs to Λ_i for some i in Ω' and so $\Lambda' = \cup \{\Lambda_i \mid i \in \Omega'\}$. Any nonzero element a of A_i can be written in the form $a = b_{j(1)} + \ldots + b_{j(t)}$ for some elements $j(1), \ldots, j(t)$ of Λ. Then $a \in \ker(\alpha_j)$ for any element j of Λ not among these elements. This shows the set Λ_i is finite for each element i of Ω. But the set Ω' is assumed to be infinite and so, by the remark before this observation, there exists an injective function from Λ' to Ω'. A similar argument with the roles of Λ' and Ω' reversed shows that there exists an injective function from Ω' to Λ'. By the Schroeder–Bernstein Theorem we now conclude that there is a bijective function θ from Ω' to Λ'. $\square$

EXERCISES

1. Suppose that B is an injective submodule of an injective left R–module A and that there exists an R–monomorphism α from A to B. Show that A and B are isomorphic.

2. Use the result in Exercise 1 to show that if A and B are left R–modules each of which is isomorphic to a submodule of the other, then the injective hulls of A and B are isomorphic.

§4 FOR PRINCIPAL IDEAL DOMAINS

We will now illustrate the program described in the Section §2 by cataloging the uniform injective modules over a principal ideal domain R. In order to discuss the ideals of R, we first choose a representative set P of primes for R. Thus for each prime p in R there is a *unique* p' in P such that p divides p'. Each ideal is of the form Ra for some a in R. When is Ra meet–irreducible? If a = 0 then Ra = (0) is meet–irreducible by the absence of zero–divisors in R. If a is a unit, then Ra = R is meet–irreducible (but of no interest to us). Now suppose that a is neither zero nor a unit. We may assume that $a = p_1^{e(1)} \cdots p_n^{e(n)}$, where n and the exponents are positive integers and each p_i is in P. If n > 1 then Ra is not meet–irreducible because $Ra = Rp_1^{e(1)} \cap Rp_2^{e(2)} \cap \ldots \cap Rp_n^{e(n)}$. We know that each nonzero proper meet–irreducible ideal is of the form Rp^e for some p in P and some positive exponent e. Now $Rp^e \sim Rp$ since R/Rp^e contains the submodule Rp^{e-1}/Rp^e, which is isomorphic to R/Rp. Thus, every nonzero meet–irreducible ideal of R is equivalent to one of $\{Rp \mid p \in P\}$. Is it possible that $Rp \sim Rq$ for distinct elements p and q of P? If $Rp \sim Rq$ then since R/Rp and R/Rq are simple, they would have to be isomorphic. Such an isomorphism is not possible because if $1 + Rp$ corresponded under such an isomorphism with $r + Rq$ we would have $p(1 + Rp) = 0$ corresponding to $p(r + Rq) = pr + Rq \neq 0$.

Thus we have observed that $\{0\} \cup \{Rp \mid p \in P\}$ is a complete set of proper inequivalent meet–irreducible ideals of R. We need only describe the injective hull of the corresponding set of factor modules: $\{R\} \cup \{R/Rp \mid p \in P\}$. We have already described the hulls of these modules in §7.5. On recalling our results there, we can conclude our description of the uniform injective R–modules as follows:

OBSERVATION 1: For a principal ideal domain R, each nonzero uniform injective R–module is isomorphic to one and only one of the modules in $\{Q\} \cup \{R(p^\infty) \mid p \in P\}$, where Q is the field of fractions of R and P is a representative set of primes in R.

In the next section we will see that every injective module over a principal ideal domain R is a direct sum of these uniform injective modules.

We have seen that when R is a principal ideal domain, each nontrivial meet–irreducible ideal L is equivalent (under $\sim$) to precisely one ideal of the form Rp, where p is a prime.

EXERCISES

1.   Let R be a principal ideal domain and let P be a

representative set of primes in R. Show that if an element a of R is neither zero nor a unit then $Ra = Rb$, where b is a product of positive powers of members of P.

2. Let R be a principal ideal domain, p a prime element of R, and e a positive integer. Show that Rp^e is meet—irreducible.

§5 FOR NOETHERIAN RINGS

In this section the classification of the injective modules over a left noetherian ring R will be reduced to the program of classifying its uniform injective modules as described in §2. This reduction is accomplished by using two powerful tools that are available for modules over these rings. One tool is the closure of the class of injective left R—modules under coproducts, which we saw in Chapter 8. The second tool is the fact that every left R—module contains a large semiuniform submodule. This latter fact will be established in the course of the next three observations.

OBSERVATION 1. If R is a left noetherian ring then each nonzero cyclic left R—module contains a nonzero uniform submodule.

VERIFICATION: Assume that there exists a proper left ideal L of R such that $A = R/L$ contains no nonzero uniform submodule. Since A is not uniform, there exist nonzero submodules A_1 and B_1 of A such that $A_1 \oplus B_1 \subseteq A$. Since B_1 is not uniform, there exist nonzero submodules A_2 and B_2 of B_1 such that $A_2 \oplus B_2 \subseteq B_1$. Continuing in this manner, we have $A_{i+1} \oplus B_{i+1} \subseteq B_i$ for each positive integer i. Then $A \supseteq \oplus\{A_i \mid i \text{ is a positive integer}\}$ and this direct sum is K/L for some left ideal K of R containing L.

Since the (strictly infinite) direct sum $\oplus A_i$ cannot be finitely generated, neither can the left ideal K. Thus R is not left noetherian. $\square$

OBSERVATION 2. If R is a left noetherian ring then each nonzero left R-module contains a nonzero uniform cyclic submodule.

VERIFICATION: Let A be a nonzero left R-module and let B be a nonzero cyclic submodule of A. By Observation 1, B contains a nonzero uniform submodule C. Let D be any nonzero cyclic submodule of C. Since any submodule of a uniform module is uniform, D is a nonzero uniform cyclic submodule of A. $\square$

OBSERVATION 3. If R is a left noetherian ring and if A is a left R-module then the submodule B of A generated by a maximal uniformly independent set M is necessarily large in A.

VERIFICATION: Suppose that B is not large in A. Then there is a nonzero submodule C of A such that $B \oplus C \subseteq A$. By Observation 2, C contains a nonzero element c for which Rc is uniform. Then $M \cup \{c\}$ is a uniformly independent subset of A that is strictly larger than M, contradicting the assumed maximality. We therefore conclude that B is large in A. $\square$

Before going on to the major result of this section, we need one more technical observation, the verification of which will be left as an exercise. Compare this result to Observation 2 of §8.2.

OBSERVATION 4. Let A be a left R-module containing a submodule of the form $\oplus\{C_i \mid i \in \Omega\}$. For each i in Ω, let D_i be a submodule of A containing C_i as a large submodule. Then the sum ΣD_i is also direct.

We are now ready for our major result, the characterization of left noetherian rings in terms of the structure of injective modules over them. You should note that this fascinating and rather surprising result is composed of two rather unequal parts: the statement that if R is left noetherian then every injective left R-module is semiuniform, which is an easy consequence of the module—theoretic observations we have made so far in this section, and the converse statement that this property in fact characterizes left noetherian rings, which is much more difficult. In actual practice the first (easy) part is the most important and useful of the two. Anybody working with modules over left noetherian rings uses it again and again. The second part serves mainly as a warning that once you venture away from left noetherian rings you cannot take this valuable tool with you.

THEOREM 3. A ring R is left noetherian if and only if every injective left R-module is semiuniform.

PROOF: Let R be left noetherian and let A be an injective left R-module. Let M be a maximal uniformly—independent subset of A. For each m in M let A_m be an injective hull of Rm contained in A. Since each A_m contains the uniform module Rm as a large submodule, each A_m is uniform. From the directness of the sum $\oplus\{Rm \mid m \in M\}$ we infer the directness of $\oplus\{A_m \mid m \in M\}$ by Observation 4. From the maximality of M and Observation 3 we conclude that $\oplus Rm$ is large in A. Then $\oplus A_m$ is certainly also large in A and, since R is left noetherian, it is also injective. Thus it must coincide with A and so we have proven that $A = \oplus\{A_m \mid m \in M\}$, where each A_m is uniform.

Now assume that every injective left R-module is semiuniform. Let $\{A_i \mid i \in \Omega\}$ be a set of nonisomorphic simple left R-modules such that any simple left R-module is isomorphic to one of them. For each i in Ω, select an injective hull D_i of A_i. Let $D = \amalg D_i$ and let G be a coproduct of a countably—infinite number of copies of D. If

$\{C_h \mid h$ a positive integer$\}$ is a countable set of injective hulls of simple left R-modules then $\amalg C_h$ is isomorphic to a direct summand of G and so if G is injective then so is any such coproduct $\amalg C_h$. Using Observation 3 of §8.4, we conclude that the injectivity of G implies the left noetherianness of R. Thus we want to show that G is injective.

Let E be an injective hull of G. By assumption, $E = \oplus\{E_j \mid j \in \Lambda\}$, where each E_j is a uniform injective submodule of E. For each i in Ω, let $\Lambda(i) = \{j \in \Lambda \mid E_j \cong D_i\}$. If n is a positive integer then, by repeated application of Observation 1 of §3, we see that there exist elements $j(1), ..., j(n)$ of Λ such that the coproduct of n copies of D_i is isomorphic to $E_{j(1)} \oplus ... \oplus E_{j(n)}$. This implies that the set $\Lambda(i)$ is infinite for each i in Ω. Set $\Lambda' = \cup\{\Lambda(i) \mid i \in \Omega\}$. Then G is isomorphic to a direct summand of the direct summand $\oplus\{E_j \mid j \in \Lambda'\}$ of E and so is injective.   □

A principal ideal domain is necessarily left noetherian. In §4 we cataloged the uniform injective modules over such rings. Theorem 3, together with this catalog, yields the following corollary.

COROLLARY. If R is a principal ideal domain then an R-module is injective if and only if it is isomorphic to the direct sum of copies of modules selected from the set $\{Q\} \cup \{R(p^\infty) \mid p \in P\}$, where Q is the field of fractions of R and P is a representative set of primes of R.

EXERCISES

1. Verify Observation 4.

Let R be a commutative noetherian ring. Exercises 2 through 8 will deal with R and its modules. This sequence of exercises culminates in the classification, due to Matlis [1968], of the injective modules over commutative noetherian rings.

2. An ideal H of R is *prime* if and only if for all r, r' ∈ R we have rr' ∈ H only when r ∈ H or r' ∈ H.
(a) Which ideals of the ring $\mathbb{Z}$ of integers are prime?
(b) Show that (0) is prime if and only if R is an integral domain.
(c) Show that a prime ideal of R must be meet−irreducible.

3. Let A be a left R−module and let a be an element of A.
(a) Show that Ann(a) = {r ∈ R | ra = 0} is an ideal of R.
(b) Show that the cyclic submodule Ra of A is isomorphic to R/Ann(a).
(c) Suppose that B and C are large submodules of A, that every nonzero element of B has the ideal H as its annihilator, and that every nonzero element of C has the ideal I as its annihilator. Show that if A ≠ (0) then H = I.

4. Let H be an ideal of R. Show that H is prime if and only if every nonzero element of R/H has H as its annihilator.

5. If A is a left R−module, show that there is an element x of A such that Ann(x) is not properly contained in Ann(a) for any nonzero element a in A.

6. Let A be a left R−module and let x be as in Exercise 5.
(a) Show that Ann(rx) = Ann(x) for all 0 ≠ r ∈ R.
(b) Show that Ann(x) is a prime ideal of R.

7. Let A be a uniform injective left R−module and let x be an element of A as in Exercise 5.
(a) Show that A is isomorphic to an injective hull of R/Ann(x).
(b) If H is a prime ideal of R such that A is isomorphic to an injective hull of R/H, show that H = Ann(x).

8. Verify the following assertions in connection with R:

(a) For each uniform injective left R—module A there is a unique prime ideal H of R for which A is isomorphic with an injective hull of R/H.

(b) A left R—module is injective if and only if it is isomorphic to the coproduct of a family of copies of modules isomorphic to injective hulls of modules of the form R/H, where H is a prime ideal of R.

§6 PROJECTIVE INJECTIVES AND QUASI—FROBENIUS RINGS

The basic theme of Part One was projectivity. We were able to do quite a bit with the general problem of determining the structure of projective modules. Let us return to this theme and ask the following question: what is the structure of a module that is both injective and projective? Since we have just discussed injectives over principal ideal domains, the question may seem peculiar. For principal ideal domains, the only module that is projective injective is (0). If we think back to semisimple rings, however, we see another picture altogether. Every module over a semisimple ring is both projective and injective.

Our experience with injective modules suggests that we begin by asking for the structure of uniform projective injective modules.

> **OBSERVATION** 1. The uniform projective injective left R—modules are precisely the indecomposable injective principal left R—modules.

VERIFICATION: Let A be a uniform projective injective left R—module. Since A is uniform and projective, it is isomorphic to a left ideal of R by Observation 4 of §1.2. Since A is injective, this left ideal is a summand of R. Since

A is uniform, this left ideal is indecomposable. Now let A be an indecomposable injective summand of R. Since A is injective and indecomposable, it is also uniform. As a summand of R, it is certainly also projective. $\square$

For left noetherian rings this observation extends to a structure theorem for arbitrary projective injective modules.

THEOREM 4. If R is a left noetherian ring then the projective injective left R-modules are, up to isomorphism, precisely the coproducts of indecomposable injective principal left R-modules.

PROOF: Since R is left noetherian, every coproduct of injective principal left R-modules is injective and, of course, also projective. Since R is left noetherian, each projective injective left R-module is a coproduct of uniform injective left R-modules, and each of these summands must also be projective. By Observation 1, each of these summands is therefore an indecomposable injective principal left R-module. $\square$

It is possible to display the uniform injective left ideals of a left noetherian ring in a rather neat manner. Since the class of injective left R-modules over a left noetherian ring R is closed under the formation of unions of nests, we know that R contains a maximal injective left ideal A. Thus $R = A \oplus B$ for some left ideal B of R which contains no nonzero injective submodules. Since A is injective, it is semiuniform, and so we have $A = \oplus\{A_i \mid i \in \Omega\}$. Thus $R = (\oplus A_i) \oplus B$. Since R is cyclic, it is only finitely decomposable. Thus Ω is finite and we may assume that $\Omega = \{1, ..., n\}$. So we have $R = A_1 \oplus ... \oplus A_n \oplus B$, where the A_i are uniform projective injective left ideals of R. Any nonzero uniform projective injective left R-module C is isomorphic to a left ideal of R and must, by its uniformity, be isomorphic to a submodule of some A_i. (Recall Observation 4 of §1.1.) Since the image of C in A_i is injective, we must have $C \cong A_i$ by the uniformity of A_i. Thus $R = A_1 \oplus ... \oplus A_n \oplus B$ provides a complete display (possibly with repetitions) of the isomorphism types of uniform projective injective left R–modules. Then we have the following observation.

OBSERVATION 2. If R is a left noetherian ring then, up to isomorphism, there are only finitely–many uniform projective injective left R–modules.

There is a widely–studied class of rings, the quasi–Frobenius rings, that have been characterized in many ways. A close examination of these rings would carry us in a direction tangential to the basic program of this book. However, the structure of the projective and(?) injective modules over quasi–Frobenius rings is known and is available to us instantly if we will set down the following highly unorthodox definition of a quasi–Frobenius ring. In spite of the peculiarity of the definition, it is actually equivalent to the various other definitions in common use.

DEFINITION. A ring R is *quasi-Frobenius* if the class of projective left R–modules coincides with the class of injective left R–modules.

Note that semisimple rings are quasi–Frobenius since each module over a semisimple ring is both projective and injective. It is not very difficult to verify that proper factor rings of principal ideal domains are always quasi–Frobenius. One of the stimuli to the development of the theory of quasi–Frobenius rings has been the following fact, which we will not prove: if the characteristic of a field F is a prime that divides the order of a finite group G, then the group ring $F[G]$ is not semisimple but is quasi–Frobenius.

In order to derive the structure of the projective (= injective) modules over quasi–Frobenius rings from Theorem 4, we need only two simple observations.

OBSERVATION 3. Every quasi–Frobenius ring is left noetherian.

VERIFICATION: If R is a quasi–Frobenius ring then, by definition, the class of all injective left R–modules is closed under taking coproducts. $\square$

OBSERVATION 4. If R is a quasi–Frobenius ring then every principal left R–module is injective.

VERIFICATION: If R is a quasi–Frobenius ring then it is injective as a left module over itself and hence so is any direct summand of it. $\square$

We now have the following immediate corollary of Theorem 4.

COROLLARY: If R is a quasi–Frobenius ring then the projective (= injective) left R–modules are, up to isomorphism, precisely the semiprincipal left R–modules.

Notice that as a special case of our earlier discussion we have for each quasi–Frobenius ring R a decomposition $R = A_1 \oplus \ldots \oplus A_n$, where each A_i is a uniform projective injective left R–module and every uniform projective injective left R–module is isomorphic to at least one of the A_i. The situation here strongly resembles the special case in which R is semisimple.

EXERCISES

1. Let R be the ring of integers modulo p^n, where p is a prime and n is a positive integer.
(a) Classify the uniform injective left R–modules in R.
(b) Apply Theorem 3 to show that the injective left R–modules are precisely the free left R–modules.
(c) Conclude that all projective left R–modules are free.

(d) Show that R is a quasi–Frobenius ring.
(e)   Generalize (a) through (d) to the case in which R is a
factor ring of an arbitrary principal ideal domain modulo the
ideal generated by a nonzero power of a prime element.

2. Let R be a quasi–Frobenius ring.
(a)   Show that every left R–module is isomorphic to a
submodule of a projective (free) left R–module.
(b)   Show that every left R–module is isomorphic to a factor
module of an injective left R–module.

3. Verify the following assertions concerning a ring R:
(a)   Each left R–module is isomorphic to a submodule of a
projective (free) R–module if and only if each injective left
R–module is projective.
(b)   Each left R–module is isomorphic to a factor module of an
injective left R–module if and only if each projective (free) left
R–module is injective.

4.   Show that if a ring is both left hereditary and
quasi–Frobenius then it must be semisimple.

5.   A left R–module   A   is *continuous* if and only if the
following two conditions are satisfied:
(a) Any submodule of A is a large submodule of some direct
summand of A; and
(b) Any submodule of A which is isomorphic to a direct
summand of A is itself a direct summand of A.
Show that a ring R is quasi–Frobenius if and only if every
projective left R–module is continuous.

NOTES FOR CHAPTER 9

Chapter 9 as a whole is based for the most part on the
work of Matlis [1958]. Attention was drawn to the class of

modules that are simultaneously projective and injective by Jans [1959]. Our unusual definition of quasi—Frobenius rings can be connected to the traditional definition of Nakayama [1939] by means of the work of Faith [1966] and Faith & Walker [1967]. See Kasch [1977].

Observation 2 of Section §3 is a special case of a more general result known as the Krull—Remak—Schmidt—Azumaya Theorem which does not deal with injective modules but rather with modules isomorphic to a coproduct of modules having local endomorphism rings. For proofs of this more general result, see Sharpe & Vámos [1972], on which the approach given here is based, or Anderson & Fuller [1974].

From Faith [1966] and Faith & Walker [1967] we can show that the following seven assertions are equivalent for a ring R:

(a) Every left R—module is isomorphic to a submodule of a free left R—module.

(b) Every left R—module is isomorphic to a submodule of a projective left R—module.

(c) All injective left R—modules are projective.

(d) R is quasi—Frobenius.

(e) All projective left R—modules are injective.

(f) All free left R—modules are injective.

(g) Each left R—module is isomorphic to a factor module of an injective left R—module. Further characterizations of quasi—Frobenius rings have been given by Rutter [1969], and additional such characterizations are still being discovered. See Oshiro [1984] for an example of more recent results of this sort. Refer also to Tachikawa [1973].

Part Three

COUNTABILITY

10

Finitely-Generated Modules

§1 INTRODUCTION

We will begin our considerations of finitely–generated modules by asking the following question. For an arbitrary ring R, what closure properties does the class of finitely–generated left R–modules have? Any factor module of a finitely–generated left R–module must be finitely generated. Consequently, the class of finitely–generated left R–modules is closed under taking factor modules and direct summands. The coproduct of a family of nonzero finitely–generated left R–modules will be finitely generated if and only if the family is finite. In Chapter 11 countably–infinite coproducts of finitely–generated modules will be examined and used. Rings differ with regard to the property of submodules of finitely–generated modules being necessarily finitely generated.

> **OBSERVATION** 1. A ring R has the property that each submodule of each finitely–generated left R–module is finitely generated if and only if R is left noetherian.

> **VERIFICATION:** If each submodule of each

finitely-generated left R-module is finitely generated then, in particular, each left ideal of R must be finitely generated. Now suppose that R is left noetherian. Then by Observations 1 and 2 of §8.4 we see that each submodule of each finitely-generated left R-module must be finitely-generated. □

The preceding verification of Observation 1 allowed us to review the powerful machinery of injectivity that was developed in Part Two. In Exercises 1 through 3, we have outlined a traditional proof of this observation that uses only elementary concepts. For one particular class of left noetherian rings, the structure of finitely-generated modules has long been known. In the next section we will prove this structure theorem by a method that continues our review of the theory of injective modules.

EXERCISES

1. Let A be a submodule of a left R-module B. Suppose that G is a subset of A that generates A and that H is a subset of B for which {h + A | h ∈ H} generates B/A.
(a) Show that G ∪ H generates B.
(b) Show that if both A and B/A are finitely generated, then so is B.

2. For an arbitrary ring R, a left R-module A is said to be *noetherian* if A and each of its submodules is finitely generated.
(a) Show that factor modules of noetherian left R-modules are noetherian.
(b) Show that a left R-module B must be noetherian if it contains a submodule A for which both A and B/A are noetherian.

3. Let R be a left noetherian ring.
(a) Use Exercise 2(b) and finite induction to show that each

free left R–module with a finite basis is noetherian.

(b)   Use Chapter 5 and Exercise 2(a) to show that each finitely–generated left R–module is noetherian––that is, that submodules of finitely–generated left modules over left noetherian rings are finitely generated.

4.   Let   A   be a finitely–generated left R–module having an injective hull   E   which is also projective.   Show that   E   is finitely generated.

§2   FOR PRINCIPAL IDEAL DOMAINS

We are going to embark on a discussion of the structure of finitely–generated modules over principal ideal domains.   Before we do so, however, we must first make the following observation, the proof of which will be left as an exercise.

> **OBSERVATION** 1.   If   R   is a principal ideal domain having field of fractions   F   then every nonzero finitely–generated R–submodule of   F   is isomorphic to   R.

Now we are ready to state our main result.

> **THEOREM** 5.   If   R   is a principal ideal domain then a nonzero R–module   A   is finitely generated if and only if it is a direct sum of a finite family of cyclic submodules each of which is isomorphic either to   R   or to R/Rp^n for some prime   p   of   R   and some positive integer n.

VERIFICATION: One direction is trivial: if   A   is a direct sum of a finite family of cyclic submodules then it is certainly finitely generated.   The hard part is proving the converse direction.

Since   principal   ideal   domains   are   surely   noetherian,   the previous section has provided us with the following tools to be used in our discussion: A can contain no submodule that is a direct sum of an infinite family of nonzero submodules and, in particular, every independent subset of A is finite. According to the following argument, it will be sufficient to show that each such A has a direct summand of one of the specified types. If we assume that every nonzero finitely-generated R-module contains a direct summand of the desired sort, then we can decompose successively, with nonzero summands C_i of the desired type, to produce $A = C_1 \oplus B_1 = C_1 \oplus C_2 \oplus B_2 = ...$. This process must terminate in an equality $A = C_1 \oplus C_2 \oplus ... \oplus C_n$ for some positive integer n since otherwise A would contain a submodule $\oplus\{C_i \mid i \text{ a positive integer}\}$, where each C_i is nonzero. Thus, we proceed to show that A has a nonzero summand isomorphic either to R or to R/Rp^n for suitable p and n.

Let $a_1, ..., a_k$ be a maximal uniformly independent subset of A. For each i $(1 \leq i \leq k)$ let $Ra_i \subseteq H_i$ be an embedding of the cyclic submodule Ra_i of A in an injective hull. The inclusion map $\oplus Ra_i \to \amalg H_i$ may be enlarged to a map $\beta{:}A \to \amalg H_i$ by the injectivity of $\amalg H_i$. Since R is noetherian and since $\{a_i\}$ is a maximal uniformly-independent subset of A, $\oplus Ra_i$ is large in A. Since $\ker(\beta) \cap \oplus Ra_i = (0)$, it follows that β is a monomorphism. Consequently, from this point on we may regard A as a submodule of $\amalg\{H_i \mid 1 \leq i \leq k\}$, where each H_i is an injective hull of a nonzero uniform cyclic R-module. In Section §9.4 we found that the nonzero uniform cyclic R-modules are precisely the modules that are isomorphic to R or R/Rp^n for suitable p and n. In §7.5 we determined that the injective hulls of these modules are F and $R(p^\infty)$, where F is the field of fractions of R and $R(p^\infty)$ was constructed as the union of a certain nest of cyclic R-modules. Now let us consider the nature of the images, C_i, of A under the projections into the various H_i $(1 \leq i \leq k)$. Each C_i is a nonzero finitely-generated submodule of H_i. But each nonzero finitely-generated submodule of F is isomorphic to R by Observation 1 and each nonzero finitely-generated submodule of $R(p^\infty)$ is isomorphic to R/Rp^n for some nonnegative integer n by Observation 4 of §7.5. Consequently, from this point on, we may regard A as a submodule of $\amalg\{C_i \mid 1 \leq i \leq k\}$, where each C_i is either R or R/Rp^n for some prime p in R and some positive integer n and where, moreover, the projection of A onto each C_i is an epimorphism.

We have $A \subseteq C_1 \amalg \ldots \amalg C_k$ and, by rearranging the C_i if necessary, we may assume that one of the following is the case:

(1) $C_1 = R$;

(2) $C_1 = R/Rp^n$ and there is no C_i for which $C_i = R$ or $C_i = R/Rp^{n+t}$ for any positive integer t.

In case (1), the kernel of the projection from A to C_1 is $A \cap B$, where $B = C_2 \amalg \ldots \amalg C_k$. The image of this projection is $C_1 = R$, which is projective. Consequently, $A = C \oplus (A \cap B)$ for some submodule C of A isomorphic to R.

In case (2), let q be an element of R that is not divisible by p but for which $qC_i = (0)$ for all C_i not of the form R/Rp^t for any positive integer t. We will make use of the following elementary fact. Since q is not divisible by the prime p, there are elements u and v in R such that $1 = uq + vp^n$. Now choose an element $c = (x_1, \ldots, x_n)$ in A for which x_1 generates C_1. Then the first coordinate, x_1, of c can be expressed as a multiple of qx_1: $x_1 = 1x_1 = uqx_1 + vp^nx_1 = uqx_1$. Let C be the cyclic submodule of A generated by $qc = (qx_1, \ldots, qx_k)$, and let $B = C_2 \amalg \ldots \amalg C_k$. Then $C \cap B = (0)$ because any element r of R that satisfies $rqx_1 = 0$ also satisfies $rqc = (0,0, \ldots, 0)$. It follows that the projection of C into C_1 is a monomorphism, and it must also be an epimorphism since $x_1 = uqx_1$ is the image of uqc. Thus $C \cong C_1 = R/Rp^n$. To complete the discussion of case (2), we need only verify that $A = C + (A \cap B)$. To do this, observe that for any element $a = (y_1, \ldots, y_k)$ in A there is an element r in R for which $y_1 = rx_1 = ruqx_1$, and consequently $a = ruqc + (a - ruqc)$ is in $C + (A \cap B)$.

Reuniting the two cases, we have $A = C \oplus (A \cap B)$, where C is isomorphic either to R or to R/Rp^n for some prime p and some positive integer n, as desired. $\square$

If p is a prime element of a principal ideal domain R then an R-module A is p-*torsion* if and only if each nonzero element of A is annihilated by some positive power of p. We can apply Theorem 5 to completely describe the structure of finitely-generated p-torsion modules. Suppose that we are given such a module A. If $A = (0)$ then there isn't much more to be said, so we might as well assume that A is nonzero. Then, by Theorem 5, we can write A as a direct sum $A_1 \oplus ... \oplus A_n$ of cyclic submodules. Since every element of A (and so, in particular, every element of each of the A_i) is annihilated by some positive power of p, we see that each of the A_i must be of the form $R/Rp^{k(i)}$ for some positive integer $k(i)$. It certainly does no harm to rearrange the terms in this sum if necessary and assume that $k(1) \geq k(2) \geq ... \geq k(n) \geq 1$. Thus we have shown that every finitely-generated p-torsion module may be assigned a finite nonincreasing sequence of positive integers. But is this assignment unique? After all, it seems at first glance perfectly plausible that we could decompose A in some other manner as a direct sum $R/Rp^{h(1)} \oplus ... \oplus R/Rp^{h(m)}$ with $h(1) \geq h(2) \geq ... \geq h(m) \geq 1$ but with no necessary relationship between the $k(i)$ and $h(j)$. The following observation shows that this cannot, in fact, happen.

OBSERVATION 2. Let p be a prime element of a principal ideal domain R and let A be a finitely-generated p-torsion R-module which can be written as $A_1 \oplus ... \oplus A_n$ and as $B_1 \oplus ... \oplus B_m$ for submodules A_i and B_j satisfying:

(1) A_i is isomorphic to $R/Rp^{h(i)}$ for $1 \leq i \leq n$;
(2) B_j is isomorphic to $R/Rp^{k(j)}$ for $1 \leq j \leq m$;
(3) $h(1) \geq h(2) \geq ... \geq h(n)$ and $k(1) \geq k(2) \geq ... \geq k(m)$.
Then $n = m$ and $h(i) = k(i)$ for all i.

VERIFICATION: If A is a p–torsion R–module then the smallest positive integer e satisfying $p^e a = 0$ for all a in A is called the p–*exponent* of A and denoted by $e(A)$. If A has direct–sum decompositions of the above form then both $h(1)$ and $k(1)$ must equal $e(A)$ and so must equal each other. We now proceed to verify the observation by induction on the p–exponent of A.

First, assume that $e(A) = 1$. Then $h(i) = 1 = k(j)$ for all i and j. Moreover, A is a vector space over the field $R/(p)$ the dimension of which is n (if one looks at the first decomposition) or m (if one looks at the second decomposition) and so we must have $n = m$.

Now assume, inductively, that $e(A) \geq 2$ and that the observation has been verified for all finitely–generated p–torsion R–modules having p–exponent less than $e(A)$. There are integers $n' \leq n$ and $m' \leq m$ such that $h(i) \geq 2$ if and only if $i \leq n'$ and $k(j) \geq 2$ if and only if $j \leq m'$. Also, we note that $A' = \{pa \mid a \in A\}$ is a submodule of A which can be written as $R/Rp^{h(1)-1} \oplus \dots \oplus R/Rp^{h(n')-1}$ and as $R/Rp^{k(1)-1} \oplus \dots \oplus R/Rp^{k(m')-1}$. Moreover, $e(A') < e(A)$ so, by the induction hypothesis, $n' = m'$ and $h(i) = k(i)$ for all $1 \leq i \leq n'$. Since A/A' is just a vector space over $R/(p)$ the dimension of which is n (if we look at the first decomposition) and m (if we look at the second) we see that $n = m$ and $h(i) = 1 = k(i)$ for all $n' < i \leq n$. $\square$

Now let us return to an arbitrary nonzero finitely–generated R–module A. By Theorem 5, we see that either A is free or there exist prime elements $p_1, \dots, p_t$ of R such that $A = A_0 \oplus \dots \oplus A_t$ where A_0 is either a free submodule of A or is (0) and where each A_i is a p_i–torsion submodule of A the structure of which is completely prescribed, as shown in Observation 2, by a finite nonincreasing sequence of positive integers.

EXERCISES

1. Let R be a principal ideal domain having field of fractions F, and regard F as an R–module.
(a)   Show that each nonzero cyclic submodule of F is isomorphic to R.
(b) Show that, for elements a/b and c/d of F (a, b, c, d ∈ R) the module R(a/b) + R(c/d) is R(g/bd), where g is the greatest common divisor of the elements ad and bc of R.
(c) Use (a), (b), and finite induction to verify Observation 1.

2. Let R be a principal ideal domain that has (essentially) only one prime; that is, each prime of R divides each other prime of R. (The ring of rational numbers expressible with odd denominators is such a ring.) Reread the verification of Theorem 5 and find those points in the discussion that can be simplified for such a ring.

NOTES FOR CHAPTER 10

The results proven in this chapter are very old and are proven in many texts and even in some introductory texts in modern algebra. For this reason, we have felt free to give unusual proofs that are likely to be new to the readers. For more traditional proofs, see Hu [1965], Rotman [1965], and Schreier and Sperner [1959]. In particular, you may want to refer to the proof of Theorem 5, based on a technique of Rado, given by Wong [1962]. The verification of Observation 2 in §2 is based on the elegant argument in Hilton & Wu [1974].

11

Countably-Generated Modules

§1 INTRODUCTION

What are the closure properties of the class of countably-generated left R-modules? Since a homomorphic image of a countably-generated left R-module must be countably-generated, the class of countably-generated left R-modules is closed under factor modules and direct summands. The coproduct of a family of nonzero countably-generated left R-modules will be countably generated if and only if the family is at most countably infinite. Direct sums of arbitrary families of countably-generated left R-modules will be studied in Chapter 12, and the results applied in Chapter 13. Rings differ with regard to the property of submodules of countably-generated left R-modules being necessarily countably generated.

> **OBSERVATION** 1. A ring R has the property that each submodule of each countably-generated left R-module is countably generated if and only if each left ideal of R is countably generated.

Since we will not be using this observation, we will leave its proof as an exercise. Probably the simplest countably—generated left R—modules are coproducts of countable families of finitely—generated left R—modules. We will study these modules in the next section.

EXERCISES

1. Let R be a ring.

(a) If A_1 and A_2 are left R—modules each having the property that all of their submodules are countably—generated, show that $A_1 \amalg A_2$ also has this property.

(b) Let B be a submodule of a coproduct $\amalg\{A_i \mid i$ a positive integer$\}$. Show that $B = \cup\{B \cap [A_1 \amalg ... \amalg A_k \mid k$ a positive integer$\}$.

(c) Show that if each left ideal of R is countably generated then each submodule of each free left R—module with a countable basis is countably generated.

2. Use Exercise 1 to verify Observation 1.

§2 COPRODUCTS OF FINITELY—GENERATED MODULES

The countably—generated modules that are direct sums of finitely—generated submodules are easily characterized.

OBSERVATION 1. A countably—generated left R—module A is the direct sum of finitely—generated submodules if and only if each finitely—generated submodule of A is contained in a finitely—generated summand.

VERIFICATION: If   A   is   the   direct   sum   of finitely–generated submodules then the stated condition is clearly satisfied. Conversely, suppose that the containment condition is satisfied for a countably–generated left R–module A. If A is finitely–generated, no verification is needed. Therefore assume that this is not the case. Thus we can find a countable sequence a_1, a_2, ... of elements of A such that A is generated by $\{a_1, a_2, ...\}$ but by no finite subset of this countably–infinite set. Let A_1 be a finitely–generated direct summand of A which contains Ra_1, and let $A = A_1 \oplus C_1$. Let B_2 be a finitely–generated direct summand of A that contains $Ra_2 + A_1$, and let $A = B_2 \oplus C_2$. Since $A_1 \subseteq B_2$ and A_1 is a direct summand of A, A_1 is also a direct summand of B_2. Let $B_2 = A_1 \oplus A_2$. Then also $A = A_1 \oplus A_2 \oplus C_2$, and a_1, $a_2 \in A_1 \oplus A_2$. This process can be continued. Thus, there is a finitely–generated direct summand B_3 of A containing $A_1 + A_2 + Ra_3$. This yields decompositions $A = B_3 \oplus C_3$, $B_3 = A_1 \oplus A_2 \oplus A_3$, and $A = A_1 \oplus A_2 \oplus A_3 \oplus C_3$. Continuation yields a submodule $\oplus\{A_i \mid i$ a positive integer$\}$ of A. For each of the generators a_j, we have $a_j \in A_1 \oplus ... \oplus A_j$, and consequently $A = \oplus A_i$. Finally, each A_i is finitely generated since it is a homomorphic image (direct summand) of the finitely–generated left R–module B_i. $\square$

If we recast the previous observation only slightly, so that it will refer to singleton subsets rather than to finitely–generated submodules, we will produce a more effective tool.

OBSERVATION 2. A countably–generated left R–module A is the direct sum of finitely–generated submodules if, for each direct summand B of A, each element of B is contained in a finitely–generated direct summand of B.

VERIFICATION: The verification of this observation is similar to that of Observation 1, with some slight changes. We make the decomposition $A = A_1 \oplus C_1$ as before. The decomposition $A = A_1 \oplus A_2 \oplus C_2$ will be arrived at as

follows: Let $a_2 = x + y$ with x in A_1 and y in C_1. Now
choose A_2 to be a finitely-generated direct summand of C_1
containing y. Letting $C_1 = A_2 \oplus C_2$, we have $A = A_1 \oplus A_2$
$\oplus C_2$ and $a_1, a_2 \in A_1 \oplus A_2$. Continuing in this fashion will
yield, as before, $A = \oplus A_i$.   □

Observation 2 yields a powerful tool for the investigation of
projective modules.   The power of this tool will be further
enhanced once we have the Kaplansky's Theorem, which appears
in the next chapter. The tool referred to is expressed in the
following observation.

OBSERVATION 3. A ring R has the property that each
of its countably-generated projective left modules is the
direct sum of finitely-generated submodules if and only if
each element of each countably-generated projective left
R-module is contained in a finitely-generated direct
summand.

VERIFICATION: The necessity of the containment
condition is clear.   The sufficiency can be obtained from
Observation 2.   □

In the next section, we will use this observation in the
process of determining the countably-generated projective left
R-modules over one more class of rings.

§3 SEMIHEREDITARY RINGS

In Chapter 4 we introduced the broadest class of rings for
which we could reduce the structure problem for submodules of
free modules to the study of left ideals. This was the class of
left hereditary rings. The left hereditary rings belong to a larger
class for which the structure problem of arbitrary submodules of

free modules has not been reduced to the study of left ideals, but for which the structure problem for finitely–generated submodules of free modules has been so reduced. This larger class is defined as follows.

DEFINITION. A ring R is *left semihereditary* if each finitely–generated left ideal is projective.

Although we cannot handle arbitrary submodules of free modules over left semihereditary rings, those that are finitely generated are very easily handled by means of the techniques of Chapter 4.

THEOREM 6. If R is a left semihereditary ring then each finitely–generated submodule of a free left R–module is isomorphic to a direct sum of finitely–generated left ideals of R.

PROOF: Notice first that a finitely–generated submodule of a free left R–module is always contained in a finitely–generated free submodule. Thus, we begin by assuming that A is a finitely–generated submodule of a free R–module $R_1 \amalg \ldots \amalg R_n$. If $n = 1$, then A is a finitely–generated left ideal of $R = R_1$. Assume that the theorem holds for all finitely–generated submodules of free left R–modules having bases of $n-1$ elements. Let π be the projection of A into R_1. Then $\text{im}(\pi)$ is a finitely–generated left ideal of R and is therefore projective. Consequently, $A = A_1 \oplus \ker(\pi)$, where $A_1 \cong \text{im}(\pi)$. Since $\ker(\pi) \subseteq R_2 \oplus \ldots \oplus R_n$, we see that $\ker(\pi)$ is isomorphic to a direct sum of finitely–generated left ideals of R by the induction hypothesis. Thus A is isomorphic to a direct sum of finitely–generated left ideals of R, and by finite induction the proof is complete. $\square$

This theorem has five corollaries. The first four are immediate and the fifth uses Observation 1 of §4.4. The second and third corollaries are characterizations of left semihereditary rings.

COROLLARY 1. If R is a left semihereditary ring then a finitely–generated projective left R–module is isomorphic to a direct sum of finitely–generated left ideals of R.

COROLLARY 2. A ring is left semihereditary if and only if each finitely–generated submodule of each of its free modules is projective.

COROLLARY 3. A ring is left semihereditary if and only if each finitely–generated submodule of each of its projective modules is projective.

COROLLARY 4. A ring R has the property that each finitely–generated submodule of each free left R–module is free if and only if each finitely–generated left ideal of R is free.

COROLLARY 5. A commutative ring R has the property that each finitely–generated submodule of each free left R–module is free if and only if it is an integral domain in which each finitely–generated ideal is cyclic.

You are now prepared for the main result of this chapter––the theorem that justifies the introduction of semihereditary rings.

THEOREM 7. If R is a left semihereditary ring then each countably–generated projective left R–module is isomorphic to a direct sum of finitely–generated left ideals of R.

PROOF: By Observation 3 of §2 and by Corollary 1, to prove this theorem it is enough to show that each element of each projective left R–module is contained in a finitely–generated direct summand. Thus, we will assume that A is a projective left R—module and that   a   is an arbitrary element of   A.   It is remarkably easy to produce the required finitely—generated summand of   A, containing   a,   once   A   has been embedded as a direct summand in a free left R—module: since   A   is projective, there is a free left R—module   F   for which   F = A ⊕ B.   Relative to any basis of   F,   a   has only a finite number of nonzero coefficients; hence, there is a decomposition $F = F_1 \oplus F_2$, where F_1 and F_2 are free and where F_1 is finitely generated and contains   a.   Thus we have $a \in F_1 \cap A$,   and we will show that $F_1 \cap A$ is a finitely—generated direct summand of   F   and consequently also of   A.   Let π be the projection of F_1 onto B   with respect to the decomposition   F = A ⊕ B.   The kernel of π is $F_1 \cap$ A   and the image of π is a finitely—generated submodule of the projective module   B.   By Corollary 3, $im(\pi)$ is projective, and consequently $F_1 = (F_1 \cap A) \oplus C$ for some submodule   C of F_1 isomorphic to $im(\pi)$.   Then $F_1 \cap A$ is finitely generated since it is a homomorphic image (in fact a direct summand) of F_1,   and $F = F_1 \oplus F_2 = (F_1 \cap A) \oplus C \oplus F_2$ shows that $F_1 \cap A$ is a direct summand of   F   and therefore also of   A,   as required.   □

Perhaps the most transparent examples of semihereditary rings that are not hereditary are found among the rings of sets that constitute the subject of the next section.   In order to study them, we will need to make an additional observation on semihereditary rings, the verification of which will be left as an exercise.

OBSERVATION 1.   If   R   is a ring having the property that every cyclic left ideal is a direct summand then   R   is semihereditary.   Moreover, every countably—generated left ideal of   R   is semiprincipal.

EXERCISES

1.   Let   A   be a left R—module having the property that each of

its cyclic submodules is a direct summand.

(a) Show that if $A = B \oplus C$ and $a = b + c$ for elements a
in A, b in B, and c in C, then $A = B \oplus Rc \oplus D$ for
some submodule D of C.

(b) Show that each finitely–generated submodule of A is a
direct summand of A.

(c) Show that each countably–generated submodule of A is a
direct sum of cyclic summands of A.

2. Verify Observation 1.

3. Show that a ring which is both left semihereditary and
quasi–Frobenius is semisimple.

§4 RINGS OF SETS

Let U be an arbitrary set and let P(U) be the power set
of U (that is, P(U) is the set of all subsets of U). It is
easy to verify that P(U) is a commutative ring if we define A
+ B = $(A \cup B) \setminus (A \cap B)$ and $A \cdot B = A \cap B$ for arbitrary
subsets A and B of U. The identity element of P(U) is
U itself. Each ideal L of P(U) is closed not only under
addition and multiplication but also under union, since $A \cup B$
$= A + B + A \cdot B$. The object of this section is to show that the
ring P(U) is always semihereditary but not always hereditary.
That P(U) is always semihereditary is a consequence of the
following observation, which contains extra information as well.

OBSERVATION 1. Each finitely–generated ideal of P(U)
is a cyclic direct summand of P(U).

VERIFICATION: For an arbitrary subset A of U, the
ideal of P(U) generated by A is the set P(A) of all subsets
of A and, since $P(U) = P(A) \oplus P(U \setminus A)$, every cyclic ideal

of P(U) is a direct summand. Now let L be an ideal of P(U) generated by a finite number of subsets $A_1, ..., A_n$ of U. Then L must contain $A = A_1 \cup ... \cup A_n$ and, since the ideal generated by A contains each of the A_i, we have L = P(A). Thus the finitely-generated ideals of P(U) are precisely the cyclic direct summands of P(U). $\square$

From this observation and from Theorem 7 we can arrive at a very clear description of the countably-generated projective modules over a ring of the form P(U).

OBSERVATION 2. Each countably-generated projective P(U)-module is semiprincipal, and each principal P(U)-module is of the form P(A) for some subset A of U.

If the set U is finite then P(U) is semimple and therefore certainly hereditary. However, if U is infinite then P(U) is not hereditary. Since the countably-generated ideals of P(U) are always projective by Observation 1 of §3, it is difficult at this stage to show that P(U) is not hereditary when U is not finite. However, in §13.1 we will prove an extended version of Theorem 7, namely that if R is a left semihereditary ring then every projective left R-module is isomorphic to a direct sum of finitely-generated left ideals. If we accept this result now, we conclude that:

OBSERVATION 3. Every projective P(U)-module is semiprincipal.

This will allow us to obtain the following result.

OBSERVATION 4. If U is an uncountable set then the ideal I of P(U) which consists of all countable subsets of U is not projective.

VERIFICATION: Suppose that I is projective. Then it

must be semicyclic, and there is a family $\{A_i \mid i \in \Omega\}$ of countable subsets of U for which $I = \oplus\{P(A_i) \mid i \in \Omega\}$. By the directness of the decomposition, distinct pairs chosen from $\{A_i \mid i \in \Omega\}$ are disjoint. Since each singleton subset of U is in I, we see that $\cup\{A_i \mid i \in \Omega\} = U$. Thus $\{A_i \mid i \in \Omega\}$ is a partition of U into countable subsets. The index set Ω must thus be uncountable. Now let Λ be a countably-infinite subset of Ω, and let $A = \cup\{A_i \mid i \in \Lambda\}$. Since A is countable, it is in I. By the construction of A, it cannot be expressed as the union of a finite number of subsets from the various A_i. Thus, the direct decomposition of I is contradicted, and we conclude that I cannot be projective. $\square$

The rings of the form $P(U)$ constitute a special subclass of the class of boolean rings.

DEFINITION. A ring R is *boolean* if $r^2 = r$ for all r in R.

OBSERVATION 5. Boolean rings are semihereditary.

Boolean rings are in turn a subclass of the class of (Von Neumann) regular rings.

DEFINITION. A ring R is *regular* (in the sense of Von Neumann) if for each r in R there exists an element r' in R satisfying $rr'r = r$.

OBSERVATION 6. Regular rings are semihereditary.

EXERCISES

1. Verify Observation 5.

2. Suppose that R is a ring containing elements r and r' for which $rr'r = r$. Show that $r'r$ is an idempotent element of R and that $Rr = Rr'r$.

3. Show that a ring is regular if and only if each of its cyclic left ideals is a direct summand.

4. Verify Observation 6.

5. Let U be a finite nonempty set having n elements. Show that $P(U)$ is isomorphic to the ring direct product of n copies of the field of integers modulo 2.

6. Let U be a nonempty set.
(a) Show that distinct cyclic ideals of $P(U)$ cannot be isomorphic as left $P(U)$–modules.
(b) With each finite (indexed) nest $A_1 \subseteq A_2 \subseteq ... \subseteq A_n$ of subsets of U, associate the left $P(U)$–module $P(A_1) \amalg P(A_2) \amalg ... \amalg P(A_n)$. Show that each finitely–generated projective left $P(U)$–module is isomorphic to the module associated with precisely one such nest.

7. Let R be a divison ring and let A be a left R–module. Show that the ring of endomorphisms of A is regular.

8. Let R be a boolean ring.
(a) Show that $x + x = 0$ for every element x in R.
(b) Show that R is commutative.
(c) If I is a maximal ideal of R, show that R/I is isomorphic to the field $\mathbf{Z}/(2)$ of integers modulo 2.

9. Let R be a boolean ring and let r be a nonzero element of R.
(a) Show that r is not contained in the ideal $R(1-r)$.
(b) Let $\mathcal{A}$ be the set of all ideals of R containing $R(1-r)$ but not containing r. Let $\mathcal{M}$ be a maximal nest in $\mathcal{A}$, and let $K = \cup\mathcal{M}$. Show that K is a maximal ideal of R that does not contain r.

10. Let R be a boolean ring.

(a) Use Exercise 9 to observe that each nonzero element of R lies outside at least one maximal ideal of R.

(b) Let a and b be distinct elements of R and let I be a maximal ideal of R that does not contain a + b. Show that precisely one of a, b belongs to I.

(c) Let U be the set of all maximal ideals of R. Let φ be the function from R to P(U) defined by $\varphi(a) = \{I \in U \mid a \notin I\}$. Show that φ defines an isomorphism of the ring R with the image of φ in P(U).

11. Show that a ring which is both regular and left noetherian is semisimple.

12. Let R be a ring.

(a) Show that R is regular if and only if every cyclic left ideal of R is a direct summand.

(b) Show that R is regular if and only if every finitely–generated left ideal of R is a direct summand.

(c) If R is regular, show that any factor ring of R is also regular.

NOTES FOR CHAPTER 11

Observation 3 of §2 is due to Kaplansky [1958]. Theorem 6 appears in Cartan & Eilenberg [1956]. Theorem 7 (and its extended version, which appears in §13.1) belongs to Albrecht [1961]. As a guide to further results related to Theorem 7, see Cohn [1985].

Rings regular in the sense of Von Neumann are interesting not only in their own right but also have important applications to functional analysis and other fields of mathematics. The best source for further information about such rings is Goodearl [1979].

Exercise 10 of §11.4 culminates in the representation theorem for boolean rings due to M. Stone [1936].

12

Coproducts of Countably-Generated Modules

§1 INTRODUCTION

In his study of the structure of projective modules, Kaplansky [1958] found that every projective left R−module must be the direct sum of countably−generated submodules. He obtained this fundamental result as a corollary of a theorem that is the subject of the present chapter: the class of coproducts of countably−generated left R−modules is closed under taking direct summands. The present chapter is an exposition of Kaplansky's proof of this closure property, and the next chapter consists of structural consequences for projective modules that are derived through the use of this closure property.

If we wish, we may regard Kaplansky's closure theorem as a structure theorem in its own right--that is, as a structure theorem for direct summands of coproducts of countably−generated modules. However, since we will not develop any further structural insight into arbitrary coproducts of countably−generated modules, the fundamental role played by the closure theorem in our study is that of a very powerful lemma for the investigation of projective modules.

§2   KAPLANSKY'S THEOREM

The purpose of this section is to prove that a direct summand of a coproduct of countably–generated left R–modules is again a coproduct of countably–generated left R–modules. We begin with the following setting. Let R be an arbitrary ring, and let M be a left R–module for which we have two decompositions: $M = A \oplus B$ and $M = \oplus\{M_i \mid i \in \Omega\}$, where each M_i is countably–generated. We ask what the second decomposition can tell us about A. The approach we take to this question will lead us to consider nests of subsets of Ω. We can describe the construction of these nests most clearly after introducing (purely for expository purposes) the concept of an *ordinal embedding*. The next paragraph is a description of this concept. Following the notation of §11.4, the set of all subsets of Ω is denoted by $P(\Omega)$.

We will use the fact that an ordinal number may be identified with the set of all ordinal numbers that precede it. By an *embedding* of an ordinal k in $P(\Omega)$, we mean a function f from k to $P(\Omega)$ that is one–to–one and preserves all unions (or at least upper bounds) that exist in k. This requirement that f preserve unions can be interpreted very easily by describing the effect of this requirement on the nature of the image im(f) of f. The indexed family of subsets im(f) = $\{f(h) \mid 0 \leq h < k\}$ is a nest for which f(h) is a proper subset of f(h') whenever $h < h'$ and for which $f(h') = \cup\{f(h) \mid 0 \leq h < h'\}$ whenever h' is a limit ordinal preceding k. Each ordinal embedding $f:k \to P(\Omega)$ determines a nest of direct summands $M_h = \oplus\{M_j \mid j \in f(h)\}$ of M $(0 \leq h < k)$ that will be fundamental in our discussion. (To prevent ambiguity in the meaning of M_h, we assume that the index set Ω contains no ordinal numbers.) We will be concerned not with arbitrary ordinal embeddings but only with those that are appropriate for the work to be done with respect to the specified decomposition of M. We say that an ordinal embedding $f:k \to P(\Omega)$ is *appropriate* if three conditions are satisfied:

(1) $k \geq 1$ and $f(0)$ is the empty set;

(2) $M_h = (A \cap M_h) \oplus (B \cap M_h)$ for all $0 \leq h < k$;

(3) $f(h + 1) \setminus f(h)$ is countable for all $0 \leq h < h + 1 < k$.

Notice that when we work with appropriate ordinal embeddings we have numerous summand relationships. For $0 \leq h < k$, the module $A \cap M_h$ is a direct summand of M_h and therefore also of M and of A. If $h+1 < k$ then $A \cap M_h$ is also a direct summand of $A \cap M_{h+1}$. This multiplicity of summand relationships is at the heart of the "appropriateness" of our specially-selected ordinal embeddings. Specifically, we make the following observation.

OBSERVATION 1. If $f:k+1 \to P(\Omega)$ is an appropriate ordinal embedding then $A \cap M_k = \oplus\{N_h \mid 0 \leq h < k\}$, where each N_h is chosen such that $A \cap M_{h+1} = (A \cap M_h) \oplus N_h$.

VERIFICATION: The sum $\oplus N_h$ is direct since any violation of directness would violate the directness of the decomposition $A \cap M_{h+1} = A \cap M_h \oplus N_h$ for some $0 \leq h < k$. The containment $\oplus N_h \subseteq A \cap M_k$ is clear from the choices of the N_h. Since $M_0 = (0)$, we have $A \cap M_0 \subseteq \oplus N_h$. Suppose that $A \cap M_j \subseteq \oplus N_h$ for all $j < j' \leq k$. If j' is a limit ordinal then $A \cap M_{j'} = A \cap (\cup\{M_j \mid j < j'\}) = \cup\{A \cap M_j \mid j < j'\} \subseteq \oplus N_h$, and if j' is not a limit ordinal then $j'-1$ is defined and $A \cap M_{j'} = A \cap M_{j'-1} \oplus N_{j'-1} \subseteq \oplus N_h$. Thus, by transfinite induction, we see that $A \cap M_k \subseteq \oplus N_h$, which completes the verification. $\square$

We have some valuable knowledge about the direct summands N_h in Observation 1.

OBSERVATION 2. Referring to the direct summands N_h $(0 \leq h < k)$ in Observation 1:

(1) Each N_h is isomorphic to a direct summand of $\oplus\{M_j \mid j \in f(h+1) \setminus f(h)\}$, and

(2) Each N_h is countably generated.

VERIFICATION: Assertion (1) follows from the following sequence of isomorphisms: $\oplus\{M_j \mid j \in f(h+1) \setminus f(h)\} \cong M_{h+1}/M_h \cong [(A \cap M_{h+1}) \oplus (B \cap M_{h+1})]/[(A \cap M_h) \oplus (B \cap M_h)] \cong [(A \cap M_{h+1})/(A \cap M_h)] \oplus [(B \cap M_{h+1})/(B \cap M_h)] \cong N_h \oplus [(B \cap M_{h+1})/(B \cap M_h)]$. Assertion (2) follows from assertion (1), the countability of $f(h+1) \setminus f(h)$, and the countable generation of the M_j. □

We have observed that an appropriate ordinal embedding of the form $f:k+1 \to P(\Omega)$ provides a direct decomposition of the associated submodule $A \cap M_k$ of A. The next natural step would be to try to arrange for $A \cap M_k$ to be as big as possible. We will say that an appropriate ordinal embedding $f:k \to P(\Omega)$ is *maximal* if its graph is not properly contained in the graph of any other appropriate ordinal embedding.

OBSERVATION 3. If $f:k \to P(\Omega)$ is a maximal appropriate ordinal embedding then k is not a limit ordinal.

VERIFICATION: If k were a limit ordinal we could enlarge f to a function $f':k+1 \to P(\Omega)$ by defining $f'(k) = \cup\{f(h) \mid 0 \leq h < k\}$. A thoughtful rereading of the defining conditions of an appropriate ordinal embedding will show that f' would be an appropriate ordinal embedding properly enlarging f. □

OBSERVATION 4. There exists a maximal appropriate ordinal embedding $f:k \to P(\Omega)$.

VERIFICATION: There is always at least one appropriate

ordinal embedding, namely the function $t:1 \rightarrow P(\Omega)$ defined by setting $t(0)$ equal to the empty set. Let $\mathcal{F}$ be the family consisting of the graphs of the appropriate ordinal embeddings into $P(\Omega)$. Let $\mathcal{M}$ be a maximal nest in $\mathcal{F}$, and let G be the union of $\mathcal{M}$. If G is the graph of an appropriate ordinal embedding f then f must be maximal due to the maximality of $\mathcal{M}$. The set of first coordinates of G is a set of ordinals and is in fact an initial segment of ordinals. Thus, letting k be the least ordinal that strictly exceeds the ordinals in this initial segment, G is the graph of a function $f:k \rightarrow P(\Omega)$. We need only show that f is an appropriate ordinal embedding.

Case I: Suppose that k is not a limit ordinal. Then $k = h+1$ for some ordinal h and h must be in the domain of an appropriate ordinal embedding $f':h' \rightarrow P(\Omega)$ having its graph in $\mathcal{M}$. From the choice of k and the definition of f, it follows that $h' = k$ and $f' = f$. In particular, f is an appropriate ordinal embedding.

Case II: Suppose that k is a limit ordinal. A contradiction will follow by Observation 3 if we show that f is an appropriate ordinal embedding. Let h be any ordinal preceding k. From the construction of k and the hypothesis that k is a limit ordinal, it follows that there is an ordinal k_h for which $h+1 < k_h < k$ and an $f_h:k_h \rightarrow P(\Omega)$ having its graph in $\mathcal{M}$. The conditions required for f to be an appropriate ordinal embedding are satisfied because they are satisfied by each f_h $(0 \leqq h < k)$. $\square$

The preceding four observations will suggest how to approach the problem of proving the following theorem.

THEOREM 8. For any ring R, each direct summand of each coproduct of countably-generated left R-modules is itself a direct sum of countably-generated submodules.

PROOF: Let M be a left R-module having decompositions $M = A \oplus B$ and $M = \oplus\{M_i \mid i \in \Omega\}$, where each M_i is countably–generated. By Observations 3 and 4, there is a maximal appropriate ordinal embedding $f{:}k{+}1 \to P(\Omega)$. If we can show that $f(k) = \Omega$, then by Observation 1 we will have $A = A \cap M_k = \oplus\{N_h \mid 0 \leq h < k\}$, and we will know that each N_h is countably generated by Observation 2. Thus, we will prove the theorem in the following manner: we will assume that $f(k)$ is a proper subset of Ω, and contradict the maximality of k by constructing an appropriate ordinal embedding $g{:}k{+}2 \to P(\Omega)$ that enlarges f.

We will begin by locating the single source of difficulty that arises in the construction of an appropriate ordinal embedding $g{:}k{+}2 \to P(\Omega)$ that enlarges f. We must choose a nonempty countable subset Λ of Ω, not contained in $\mathrm{im}(f)$, in such a manner that $g(k{+}1) = f(k) \cup \Lambda$, together with $g(h) = f(h)$ for all $0 \leq h \leq k$, will define an appropriate ordinal embedding. A review of the definitions of ordinal embedding and of appropriateness will show that any such Λ will be satisfactory to meet all requirements save one, and that one is $M_{k+1} = (A \cap M_{k+1}) \oplus (B \cap M_{k+1})$. The procedure will be to choose an arbitrary j in $\Omega \setminus f(k)$ and then build Λ by adjoining additional elements of Ω to $\{j\}$ as needed to insure that $M_{k+1} = (A \cap M_{k+1}) \oplus (B \cap M_{k+1})$. Fortunately, it will turn out that only a countable subset of Ω needs to be adjoined to $\{j\}$ to produce this decomposition, and thus the theorem will follow.

Choose an arbitrary element j from $\Omega \setminus f(k)$. We will build an infinite matrix $[x_{u,v}]$ consisting of elements of M by choosing rows

$$x_{i,1} \quad x_{i,2} \quad x_{i,3} \quad \ldots$$

in succession. The first row is chosen to be a sequence of elements that forms a generating set for M_j.

The second row is chosen as follows. Let $x_{1,1} = a + b$, with $a \in A$ and $b \in B$. With respect to the decomposition $M = \oplus M_i$, we have $a = \Sigma a_i$ and $b = \Sigma b_i$, where $a_i = b_i = 0$ for all but finitely–many i in Ω. The set $\Lambda_{1,1} = \{i \in \Omega \mid a_i \neq 0 \text{ or } b_i \neq 0\}$ is finite. The second row is chosen to be a sequence of elements that forms a generating set for $\oplus\{M_i \mid i$

$\in \Lambda_{1,1}\}$.

The third row is chosen as follows. Let $x_{1,2} = a + b$, $a = \Sigma a_i$, and $b = \Sigma b_i$, where $a \in A$, $b \in B$, $a_i \in M_i$, and $b_i \in M_i$ as before. Let $\Lambda_{1,2} = \{i \in \Omega \mid a_i \neq 0$ or $b_i \neq 0\}$. The third row is chosen to be a sequence of elements that forms a generating set for $\oplus\{M_i \mid i \in \Lambda_{1,2}\}$.

In choosing the fourth row, we proceed in a similar manner. We do not, however, work with $x_{1,3}$ but with $x_{2,1}$. The pattern will be to proceed down diagonals. At this step, we produce a finite subset $\Lambda_{2,1}$ of Ω.

The remaining rows are chosen in the same manner. The succession of elements treated is $x_{1,1}, x_{1,2}, x_{2,1}, x_{1,3}, x_{2,2}, x_{3,1}, \dots$, and the corresponding finite subsets of Ω produced are $\Lambda_{1,1}, \Lambda_{1,2}, \Lambda_{2,1}, \Lambda_{1,3}, \Lambda_{2,2}, \Lambda_{3,1}, \dots$.

We are now ready to define Λ and complete the proof. Choose $\Lambda = \{j\} \cup (\cup\{\Lambda_{u,v} \mid u, v \text{ positive integers}\})$. With this choice made, the ordinal embedding $g:k+1 \to P(\Omega)$ is specified (using $g(k+1) = f(k) \cup \Lambda$), and we proceed to verify that in fact $M_{k+1} = (A \cap M_{k+1}) \oplus (B \cap M_{k+1})$. Our basic tool will be the pattern of construction of our infinite matrix and the fact that the elements of the matrix generate $\oplus\{M_i \mid i \in \Lambda\}$. Since $M_{k+1} = M_k + (\oplus\{M_i \mid i \in \Lambda\})$ and $M_k = (A \cap M_k) \oplus (B \cap M_k)$, we need only verify that for each x in $\oplus\{M_i \mid i \in \Lambda\}$ we have $a, b \in M_{k+1}$, where $x = a + b$ and $a \in A$, $b \in B$. To make this last verification, it is enough to consider the special case in which x is one of the elements of our generating set $\{x_{u,v} \mid u, v \text{ positive integers}\}$ of $\oplus\{M_i \mid i \in \Lambda\}$. Thus, consider $x_{u,v} = a + b$ and ask if a and b are in M_{k+1} as required. Because of the use of the diagonal pattern in the construction of the matrix $[x_{u,v}]$, at some nth step $(n \leq [u + v - 1]^2)$ the decomposition $x_{u,v} = a + b$ is taken up, and the submodule generated by the row chosen at this step contains both a and b. In fact, a and b both belong to $\oplus\{M_t \mid t \in \Lambda_{u,v}\} \subseteq M_{k+1}$, as required. This completes the proof of the theorem. □

Some interesting generalizations of Kaplansky's Theorem have been proven and applied in the study of algebraic questions that are outside the scope of this book. Carol Walker [1966] has given the following generalization: For a ring R and an arbitrary infinite cardinal k, a summand of a coproduct of left R—modules, each of which has a generating set of cardinality not greater than k, is itself a direct sum of submodules, each of which has a generating set of cardinality not greater than k. A generalization that moves into the setting of nonabelian groups has been given by Paul Hill [1970].

13

Projective Modules

§1 INTRODUCTION

The two previous chapters have left us in a pleasant position. We can draw some substantial conclusions about the structure of projective modules with virtually no effort. Since a projective module is a direct summand of a free module and a free module is certainly a coproduct of countably–generated modules (even cyclic modules), Theorem 8 yields the following result.

THEOREM 9. For any ring R, every projective left R–module is a direct sum of countably–generated submodules.

By this theorem, the problem of determining the structure of the projective left R–modules for some ring R is reduced to the special case of countably–generated projectives. For some rings, Observation 3 of §11.2 will then allow this problem to be reduced to the finitely–generated case. Specifically, Theorem 9 allows for an immediate improvement of Theorem 7.

THEOREM 7 (completed): If R is a left semihereditary ring then every projective left R-module is isomorphic to a coproduct of finitely-generated left ideals of R.

Does it seem likely that a theorem on the structure of modules over left semihereditary rings would give us new insight into the internal features of left hereditary rings? The following corollaries are immediate consequences of Theorem 7. Other consequences are indicated in the exercises.

COROLLARY 1. Each left ideal of a left hereditary ring R is a direct sum of finitely-generated left ideals.

COROLLARY 2. A hereditary integral domain is noetherian.

We can make use of Theorem 9 to get an even stronger result for modules which are both projective and injective. First, however, we need to make an additional observation.

OBSERVATION 1. Let A be a finitely-generated left R-module having an injective hull C which is also projective. Then C is finitely-generated.

VERIFICATION: Since C is projective, there is a monomorphism α from C to a free left R-module F. Since A is finitely-generated, there is a finitely-generated free direct summand F' of F containing $A\alpha$. Let $\pi:F \to F'$ be the projection of F onto F'. Then $A \cap \ker(\alpha\pi) = (0)$ and so, since A is large in C, it follows that $\ker(\alpha\pi) = (0)$ and hence C is isomorphic to a submodule of F'. Since C is injective, it is isomorphic to a direct summand of F' and so is finitely-generated. $\square$

We can now conclude the following.

THEOREM 10. If A is a left R-module which is both

projective and injective then it is a direct sum of finitely–generated submodules.

PROOF: By Theorem 9 it suffices to prove the assertion for the case that A is countably–generated. Observation 2 of §11.2 provides the tool we need. Let B be an arbitrary direct summand of A and let x be an element of B. We need only show that x is contained in a finitely–generated direct summand of B. Since A is injective, the same is true for B. Consequently, B contains an injective hull C of Rx. But C is also projective since it is a direct summand of A. From Observation 1 it follows that C is finitely generated. Thus C is the desired finitely–generated direct summand of B containing x. □

The remainder of this chapter is devoted to determining the structure of projective modules over the class of local rings.

EXERCISES

1. Let U be a nonempty set.
(a) Show that if A is a proper projective ideal of P(U) and A is not of the form P(U \ {u}) for any u in U, then A is properly contained in a proper projective ideal of P(U).
(b) Let W be the set of all finite subsets of U. Verify that W is an ideal of P(U).
(c) Assume now that U is infinite. Show that no maximal ideal of P(U) that contains W can be projective.
(d) Show that P(U) is hereditary if and only if U is finite.

2. A function f from the set $\mathbb{C}$ of complex numbers to itself is an *entire function* if it possesses a derivative at each z in $\mathbb{C}$.
(a) Verify that the set E of all entire functions forms a commutative ring with respect to the operations $(f + g)(z) = f(z) + g(z)$ and $(f \cdot g)(z) = f(z)g(z)$.

(b) When are two cyclic ideals of E isomorphic?

(c) Describe the finitely–generated ideals of E. (See [Helmer, 1940] for the solution.)

(d) Show that E is semihereditary.

(e) Describe the projective E–modules.

(f) Show that E is neither hereditary nor regular.

§2 LOCAL RINGS

In a division ring the zero element is the only element that is not a unit, that is, does not have a two–sided multiplicative inverse. Thus, in a very trivial way, a division ring has the property that its set of noninvertible elements is a two–sided ideal. A number of other rings have this property. For each prime integer p and each positive integer n, the noninvertible elements of the ring $\mathbb{Z}/\mathbb{Z}p^n$ constitute the principal ideal generated by the coset $p + \mathbb{Z}p^n$. For the ring consisting of those rational numbers that are expressible with odd denominators, the noninvertible elements constitute the principal ideal generated by 2. Additional examples of rings having this property are included in the exercises. These examples motivate having another look at those rings having the property that their noninvertible elements form a two–sided ideal, namely the local rings which we introduced briefly in §1.2.

Let R be a local ring, and let L be the ideal of all noninvertible elements of R. Then any left (or right) ideal of R that is proper must consist of noninvertible elements. Thus L is the unique maximal left ideal of R, the unique maximal right ideal of R, and the unique maximal two–sided ideal of R. Since L is the *unique* maximal left ideal of R, there is, up to isomorphism, only one simple left R–module––namely R/L. Since L is a two–sided ideal, R/L is a ring. From the simplicity of the *module* R/L it follows that the *ring* R/L is a division ring. By Observation 4 of §2.4, any two bases of the same free R/L-module must have the same cardinality. Observation 6 of §2.4 now applies to give one of the pleasant features of local rings:

OBSERVATION 1. Any two bases of a free module over a local ring must have the same cardinality.

We may think of the local ring R as a close relative of this division ring R/L. For an arbitrary left R-module A, it may be appropriate to regard A as a similarly close relative of the associated left R/L-module A/LA. That this intuition is approrpriate for projective modules is shown by the following theorem of Kaplansky.

THEOREM 11. Each projective module over a local ring is free.

PROOF: By Theorem 9 we may restrict our attention to countably-generated projectives. By Observation 3 of §11.2 and a brief re-examination of the direct sum decomposition made in Observation 2 of that section, we see that to prove the theorem it suffices to demonstrate that if R is a local ring then each element x of each countably-generated projective left R-module P is contained in a finitely-generated free direct summand of P. Since P is a countably-generated projective, it is a direct summand of a countably-generated free left R-module $F = P \oplus Q$. The procedure for finding a free direct summand F' of P that contains x will be to construct a basis $\{v_i \mid i$ a positive integer$\}$ of F for which, for some n, $x = r_1 v_1 + ... + r_n v_n$ and $\{v_1, ..., v_n\} \subseteq P$. Once this is done, we can choose F' to be $Rv_1 \oplus ... \oplus Rv_n$, and F' will be a direct summand of P because it is a direct summand of F. The required basis $\{v_i\}$ will be constructed by modifying a basis $\{u_i\}$ of F, which we will choose with particular attention to x.

Let $\{u_i\}$ be a basis of F with respect to which the

representation $x = \Sigma r_i u_i$ of x has the smallest number of nonzero coefficients. Let the indexing of the basis be made so that $x = r_1 u_1 + \dots + r_n u_n$ with $r_1, \dots, r_n$ all nonero. The desired basis $\{v_i\}$ is constructed by altering only the first n elements of $\{u_i\}$. For each $1 \le i \le n$, decompose u_i according to $F = P \oplus Q$ to obtain $u_i = v_i + w_i$. Then we have $x = r_1 u_1 + \dots + r_n u_n = (r_1 v_1 + \dots + r_n v_n) + (r_1 w_1 + \dots + r_n w_n)$. Since $\{x, v_1, \dots, v_n\} \subseteq P$ and since $\{w_1, \dots, w_n\} \subseteq Q$, we in fact have $\Sigma r_i w_i = 0$ and so $x = \Sigma r_i v_i$. Now let F' be the submodule of P generated by $\{v_1, \dots, v_n\}$. Then we have $x \in F' \subseteq P$, and to complete the proof it will suffice to verify that $\{v_1, \dots, v_n, u_{n+1}, u_{n+2}, \dots\}$ is a basis of F.

First we must lay out an important tool. The choice of the indexed basis $\{u_i\}$ gives us a condition on the coefficients r_i: no r_i is a right linear combination of the remaining coefficients. To justify this assertion it is enough to consider the notationally most covenient case: r_n. Suppose that $r_n = r_1 t_1 + \dots + r_{n-1} t_{n-1}$, where the t_i are also elements of R. Then $x = r_1 u_1 + \dots + r_{n-1} u_{n-1} + (r_1 t_1 + \dots + r_{n-1} t_{n-1}) u_n = r_1(u_1 + t_1 u_n) + \dots + r_{n-1}(u_{n-1} + t_{n-1} u_n)$ is an expression for x involving only $n-1$ nonzero coefficients. Since the set $\{u_1 + t_1 u_n, \dots, u_{n-1} + t_{n-1} u_n, u_n, u_{n+1}, \dots\}$ can be mentally verified to be a basis for F, we have contradicted the choice of $\{u_i\}$. This contradiction justifies our assertion.

For each v_i $(1 \le i \le n)$ we have a unique representation $v_i = r_{i1} u_1 + \dots + r_{in} u_n + w_i$, where w_i is a linear combination of the u_j for which $j > n$. We need some information about the coefficients r_{ij}. To get it, we replace each of the v_i in the equation $r_1 v_1 + \dots + r_n v_n = r_1 u_1 + \dots + r_n u_n$ by its expansion in terms of the basis $\{u_i\}$. Equating coefficients of each u_i then gives n equations: $r_1 r_{1i} + r_2 r_{2i} + \dots + r_n r_{ni} = r_i$ $(1 \le i \le n)$. Rearranging gives $r_1 r_i + \dots + r_{i-1} r_{(i-1)i} + r_i(r_{ii} - 1) + r_{i+1} r_{(i+1)i} + \dots + r_n r_{ni} = 0$. Since no r_j is a right linear combination of the others, we conclude that the following elements of R are noninvertible: $r_i, \dots, r_{(i-1)i}, r_{ii} - 1, r_{(i+1)i}, \dots, r_{ni}$. Since this conclusion is valid for each i $(1 \le i \le n)$, we may summarize that r_{ij} is noninvertible if $i \neq j$ and $1 - r_{ii}$ is noninvertible for each i. So far our discussion has been valid for arbitrary rings. We now make our first use of the fact that R is local: since $1 = r_{ii} + (1 - r_{ii})$ and $1 - r_{ii}$ is noninvertible, we conclude that r_{ii} is a unit.

To complete the proof of the theorem it is sufficient to verify the following assertion: if R is a local ring, if $\{u_i \mid i$ a positive integer$\}$ is a basis of a free left R–module F, and if

$$v_1 = r_{11}u_1 + \ldots + r_{1n}u_n + w_1$$

(*)

$$v_n = r_{n1}u_1 + \ldots + r_{nn}u_n + w_n$$

is a system of equations for which r_{ij} is a unit if and only if $i = j$ and each w_i is a linear combination of the u_j $(j > n)$ then $\{v_1, \ldots, v_n, u_{n+1}, u_{n+2}, \ldots \}$ is also a basis of F. This will be proven by induction on n. Assume that the corresponding assertion in which n is replaced by $n-1$ is true. Since r_{nn} is a unit, it is mentally verifiable that $\{u_1, \ldots, u_{n-1}, v_n, u_{n+1}, u_{n+1}, \ldots \}$ is a basis of F. (This also verifes the assertion for $n = 1$.) In the system (*), the last equation can be solved for u_n and the result substituted in the preceding equations to produce a system

$$v_1 = s_{11}u_1 + \ldots + s_{1(n-1)}u_{n-1} + y_1$$

$$v_{n-1} = s_{(n-1)1}u_1 + \ldots + s_{(n-1)(n-1)}u_{n-1} + y_{n-1}$$

where each y_i is a linear combination of v_n and the u_j $(j > n)$. In order to employ our induction hypothesis we must show that s_{ij} is a unit if and only if $i = j$. To compute s_{ij}, let r

be the inverse of r_{nn}. Then $s_{ij} = r_{ij} - r_{in}rr_{nj}$. Since neither i nor j is equal to n, $r_{in}rr_{nj}$ is certainly noninvertible and so $r_{ij} - r_{in}rr_{nj}$ is a unit if and only if r_{ij} is a unit. Thus s_{ij} is a unit if and only if $i = j$. We can now employ the induction hypothesis and conclude that, since $\{u_1, ..., u_{n-1}, v_n, u_{n+1}, ...\}$ is a basis of F, so is $\{v_1, ..., v_{n-1}, v_n, u_{n+1}, ...\}$. This completes the proof of the theorem. $\square$

EXERCISES

1. Let R be a ring. By a *formal power series* in an indeterminate X over R we mean an expression of the form $a_0 + a_1X + a_2X^2 + ...$, where the a_i are elements of R. Such expressions will be denoted briefly by Σa_iX^i. The set of all formal power series in X over R will be denoted by $R[[X]]$. Addition and multiplication of elements of $R[[X]]$ are defined by $(\Sigma a_iX^i) + (\Sigma b_iX^i) = \Sigma(a_i+b_i)X^i$ and $(\Sigma a_iX^i)\cdot(\Sigma b_iX^i) = \Sigma c_iX^i$, where $c_i = \Sigma\{a_jb_k \mid j + k = i\}$ for all i.
(a) Verify that $R[[X]]$ is a ring under these operations.
(b) Show that an element Σa_iX^i of $R[[X]]$ is invertible if and only if a_0 is invertible.
(c) Show that $R[[X]]$ is local if and only if R is local.

2. Let R be a division ring.
(a) Show that all projective left $R[[X]]$-modules must be free.
(b) Show that submodules of free left $R[[X]]$-modules are necessarily free.

3. Let R be a division ring and let $R[[X,Y]]$ denote the ring $(R[[X]])[[Y]]$, that is, the ring of formal power series in the indeterminate Y over the ring $R[[X]]$.
(a) Show that projective left $R[[X,Y]]$-modules must be free.
(b) Show that any two bases of a free left $R[[X,Y]]$-module must have the same cardinality.
(c) Show that the left ideal of $R[[X,Y]]$ generated by $\{X, Y\}$ is not projective.

4. In Exercise 1 we denoted the elements of $R[[X]]$ by the formal expressions $\Sigma a_i X^i$ with the understanding that i denotes a *nonnegative* integer. Let us now relax this restriction and examine the formal expressions of the form $\Sigma a_i X^i$ in which i is allowed to assume arbitrary integer values. Addition of such expressions can be defined as before, but the definition of multiplication breaks down since the sum $\Sigma\{a_j b_k \mid j + k = i\}$ may involve an infinite number of nonzero terms and therefore fail to be defined. Let $R\langle X\rangle$ consist of those elements $\Sigma a_i X^i$ for which the set $\{i \mid i < 0 \text{ and } a_i \neq 0\}$ is finite. Then multiplication of such elements is defined in a manner analogous to that in $R[[X]]$.

(a) Verify that $R\langle X\rangle$ is a ring.

(b) Show that $R\langle X\rangle$ is a division ring if and only if R is a division ring.

(c) Let R be a division ring. Since $R\langle X\rangle$ contains $R[[X]]$ as a subring, it can be regarded as a left $R[[X]]$–module. Show that the $R[[X]]$–module $R\langle X\rangle$ is an injective (divisible) hull of the left $R[[X]]$–module $R[[X]]$. (Help is available in McCoy [1964].)

5. Show that a commutative ring which is both local and hereditary is a principal ideal domain.

NOTES FOR CHAPTER 13

Theorems 9 and 10 are due to Kaplansky [1958] and Theorem 7 is due to Albrecht [1961]. The dependence of Chapters 11, 12, and 13 on Kaplansky's paper can hardly be overemphasized. These three chapters are essentially an embedding of Albrecht's paper into Kaplansky's with the result being incorporated into the program of the present book.

Among the most important classes of commutative rings are

Dedekind domains. These rings may be defined in several ways; the definition most in tune with our approach here is the following: a Dedekind domain is a hereditary integral domain. We have already encountered these rings in Corollary 2 to Theorem 7. The problem of describing the structure of the projective and injective modules over a Dedekind domain is reducible to considerations involving ideals of R: by Theorem 7 every projective left R—module must be isomorphic to a coproduct of (necessarily finitely generated) left ideals of R. Moreover, since R is an integral domain, every left ideal of R must be uniform and consequently the projective left R—modules are also semiuniform. By Corollary 2, R is noetherian and consequently by Theorem 3 each injective left R—module is semiuniform. Therefore each uniform injective left R—module is isomorphic to the injective hull of R/P for a unique prime ideal P of R.

In Cartan & Eilenberg [1956] or Rotman [1970] Dedekind domains are defined in terms of the "invertibility" of ideals. In Curtis & Reiner [1962] or Zariski & Samuel [1958] you will find a proof of the observation that each ideal of a Dedekind domain is uniquely expressible as a product of prime ideals. This theorem explains the long—standing interest in Dedekind domains, which harks back to work by Dedekind on Fermat's Last Theorem. Also in Zariski & Samuel [1958] you will find the interesting result that each ideal of a Dedekind domain can actually be generated by two or fewer elements.

Part Four

RINGS AND MODULES OF QUOTIENTS

14

Torsion and Torsionfree Modules

§1 INTRODUCTION

One of the fundamental constructions of commutative ring theory is that of the classical ring of quotients at a multiplicatively–closed subset of the ring. Specifically, if R is a commutative ring and S is a nonempty multiplicatively–closed subset of R not containing 0 then one can construct the ring $S^{-1}R$ by the following sequence of steps:

(a) Define an equivalence relation $\sim$ on the set $R \times S$ by setting $(r,s) \sim (r',s')$ if and only if there exists an element s'' in S such that $s''(s'r - sr') = 0$. Denote the equivalence class of the pair (r,s) by r/s and denote the set of all such equivalence classes by $S^{-1}R$.

(b) Define addition and multiplication on $S^{-1}R$ as follows: if r/s and r'/s' are elements of $S^{-1}R$, then $r/s + r'/s' = [s'r + sr']/ss'$ and $(r/s) \cdot (r'/s') = rr'/ss'$.

Moreover, if A is a left R–module then we can construct the left $S^{-1}R$–module $S^{-1}A$ by the following sequence of steps:

(c) Define the equivalence relation $\sim$ on the set $A \times S$ by setting $(a,s) \sim (a',s')$ if and only if there exists an element s''

in S such that s"(s'a − sa') = 0. Denote the equivalence class of the pair (a,s) by a/s and denote the set of all such equivalence classes by $S^{-1}A$.

(d)   Define addition and scalar multiplication on $S^{-1}A$ as follows: if a/s and a'/s' are elements of $S^{-1}A$ then a/s + a'/s' = [s'a + sa']/ss'; if r/s is an element of $S^{-1}R$ and a'/s' is an element of $S^{-1}A$ then (r/s)·(a'/s') = ra'/ss'.

In the most common application of this construction one takes S to be R \ H for some prime ideal H of R. If R is an integral domain and S = R \ {0}, then $S^{-1}R$ is precisely the field of fractions of R.

When one checks the above construction in detail, one sees that the commutativity of the ring R is vital in making it work (see Exercise 1) and so it is not obvious at all how to extend it to arbitrary noncommutative rings. (For some noncommutative rings one can use this method; see Exercises 3 and 4.) As in many cases we have already discussed in this book, it turns out that the best way to attack this problem is not "internally"––by looking at the elements of the ring as before––but "externally"––by looking at injective and projective modules. The key to this approach lies in §7.5 where we noticed that if Q is the field of fractions of a principal ideal domain R then, considered as a left R–module, it is an injective hull of R. Perhaps rings and modules of fractions can be constructed as (submodules of) injective hulls.

This approach indeed works, though it will take us three chapters to complete it. In this chapter we begin the construction by considering what, at first glance, is a totally different problem, that of the notions of torsion and torsionfree modules. In Chapter 15 we will use these to construct modules of quotients, and finally in Chapter 16 we will construct the ring of quotients of the ring and show that we in fact obtain what we intended to obtain.

EXERCISES

1.   Let S be a nonempty multiplicatively–closed subset of a commutative ring R which does not contain 0.

(a) Verify that $S^{-1}R$, as defined above, is a ring.

(b) If A is a left R-module, verify that $S^{-1}A$, as defined above, is a left $S^{-1}R$-module.

2. Let S be a nonzero multiplicatively-closed subset of a commutative ring R which does not contain 0 and let A be a finitely-generated left R-module. Show that $S^{-1}A = (0)$ if and only if there exists an element s of S satisfying $sA = (0)$.

3. Let R be an arbitrary ring and let S be a nonempty multiplicatively-closed set of nonzero elements of R satisfying the following conditions:

(a) No element of S is a zero-divisor (that is to say, if $s \in S$ then there is no nonzero element r of R such that $rs = 0$ or $sr = 0$); and

(b) If $r \in R$ and $s \in S$ there exist elements $r' \in R$ and $s' \in S$ such that $s'r = r's$.

Show that the ring $S^{-1}R$ can be constructed by the classical method described above.

4. If R is a left noetherian ring satisfying the condition that $rr' \neq 0$ for all nonzero elements r and r' of R and if S is a nonempty multiplicatively-closed set of nonzero elements of R, show that the ring $S^{-1}R$ can be constructed by the classical method described above.

5. A commutative semihereditary integral domain is known as a *Prüfer domain*. If R is a Prüfer domain and if S is a nonempty multiplicatively-closed set of nonzero elements of R, show that $S^{-1}R$ is also a Prüfer domain.

§2 EQUIVALENT INJECTIVE MODULES

First, we are going to look at classes of injective modules which are, from a homological point of view, equivalent. These classes will form the basis for our theory of localization.

DEFINITION. Injective left R—modules E and E' are *equivalent* if each of them is isomorphic to a submodule (and hence a direct summand) of a direct product of copies of the other.

> **OBSERVATION** 1. Injective left R—modules E and E' are equivalent if and only if, for any left R—module A, there exists a nonzero map from A to E when and only when there exists one from A to E'.

VERIFICATION: Suppose that E and E' are equivalent and let α be a nonzero map from a left R—module A to E. Then there exists a family $\{B_i \mid i \in \Omega\}$ of copies of E' and an R—monomorphism λ from E to ΠB_i. Since $\text{im}(\alpha)$ is a nonzero submodule of E, there exists an h in Ω such that $\text{im}(\alpha)\pi_h \neq (0)$, where π_h is the projection of the direct product onto B_h. Thus $\alpha\lambda\pi_h$ is a nonzero map from A to E'. Reversing the roles of E and E', we see similarly that if there is a nonzero map from A to E' we must also have a nonzero map from A to E.

Now, conversely, assume this condition is satisfied. Let H = $\text{Hom}_R(E,E')$ and, for each α in H, let B_α be a copy of E'. Then there is a nonzero map β from E to $\Pi\{B_\alpha \mid \alpha \in H\}$ defined as follows: if $x \in E$ and $\alpha \in H$ then $(x\beta)(\alpha) = x\alpha$. If $K = \ker(\beta)$ then $\text{Hom}_R(K,E') = (0)$ since any nonzero map from K to E' can be enlarged to a nonzero map from E to E'. Therefore E is isomorphic to a submodule of a direct product of copies of E'. A similar argument shows that E' is isomorphic to a submodule of a direct product of copies of E, and so these injective modules are equivalent. $\square$

It is easily seen, using this observation, that any two injective cogenerators are equivalent.

Our basic tool for the construction of rings and modules of quotients will be classes of injective modules equivalent in the above sense. Surprisingly, there are not that many such equivalence classes: as we shall verify later (see Exercise 10 of §14.4), the family of equivalence classes of injective left R–modules is in one–to–one correspondence with a set, and therefore has cardinality. We would like to say that it *is* a set but, strictly speaking, that is not the case. A set cannot have elements which are proper classes. However, this result implies that there is no harm in informally considering it a set.

The following observation is important in that it shows that each equivalence class of injective left R–modules contains at least one member of a very special type, which we will need later.

OBSERVATION 2. Any injective left R–module is equivalent to a left R–module which is cyclic as a right module over its endomorphism ring.

VERIFICATION: Let E be an injective left R–module and let $\{A_i \mid i \in \Omega\}$ be a set of nonisomorphic cyclic left R–modules such that any cyclic submodule of an injective left R–module equivalent to E is isomorphic to one of the A_i. Let $A = \amalg\{A_i \mid i \in \Omega\}$ and let C be an injective hull of A.

We first assert that C is equivalent to E. Suppose that there exists a nonzero map α from a left R–module M to C. Then there exists a nonzero element m of M such that $m\alpha$ is a nonzero element of A and so there is an i in Ω such that $m\alpha\pi_i \neq 0$, where π_i is the projection of A onto A_i. But A_i is a cyclic submodule of an injective left R–module E' equivalent to E and so there exists a nonzero map from M to E'. By Observation 1 there then exists a nonzero map from M to E. Conversely, suppose that there exists a nonzero map β from M to E and let m be an element of M such that $m\beta$ is a nonzero element of E. Then $Rm\beta$ is isomorphic to A_h for some h in Ω and so there exists a nonzero map from Rm to A_h and hence from Rm to C. By injectivity, this map can be enlarged to a nonzero map from M to C. This verifies that C is equivalent to E by Observation 1.

Finally, we must show that C is cyclic as a right module over its endomorphism ring S. Let I be the subset of R consisting of all those elements r satisfying $rx = 0$ for all x in C. Then I is a two–sided ideal of R. For each x in C, let C_x be a copy of C. Then we have a map α from the left R–module R to $C' = \Pi\{C_x \mid x \in C\}$ defined by $(r\alpha)(x) = rx$ for all r in R and all x in C. The kernel of this map is precisely I and so the left R–module R/I is isomorphic to a cyclic submodule of C'. But C' is equivalent to C and hence to E and so, by the construction of A, there is some h in Ω such that R/I is isomorphic to A_h. In particular, we have an R–monomorphism $\delta{:}R/I \to A \subseteq C$.

Since I is a two–sided ideal of R, we see that R/I is a ring and that C is a left (R/I)–module with scalar multiplication defined by $(r+I)x = rx$ for all r in R and all x in C. In fact, any R–endomorphism of C is also an (R/I)–endomorphism and any (R/I)–endomorphism of C is surely an R–endomorphism. Thus we can identify the rings $\mathrm{Hom}_R(C,C)$ and $\mathrm{Hom}_{R/I}(C,C)$. We will denote them both by S.

By Observation 3 of §0.5 we see that $\mathrm{Hom}_{R/I}(R/I,C)$ is a right S–module isomorphic to C. Consider the map of right S–modules from S to $\mathrm{Hom}_{R/I}(R/I,C)$ which sends an S–endomorphism β of C to $\delta\beta$. We claim that this is an S–epimorphism. Indeed, an immediate application of Baer's criterion shows that C is injective as a left (R/I)–module since it is injective as a left R–module. Therefore, for any map γ from R/I to C there is an (R/I)–endomorphism β of C satisfying $\gamma = \delta\beta$. Therefore $\mathrm{Hom}_{R/I}(R/I,C)$ is a cyclic right S–module. Since C is isomorphic to it, it too is a cyclic right S–module, which is what we wanted to show. $\square$

§3 TORSIONFREE MODULES

If E is an injective left R—module then a left R—module A is E-*torsionfree* if and only if it is isomorphic to a submodule of an injective module equivalent to E. In other words, A is E—torsionfree if and only if it is embeddable in a direct product of copies of E.

OBSERVATION 1. If E is an injective left R—module then the class of E—torsionfree left R—modules is closed under taking submodules, direct products, isomorphic copies, and injective hulls. Moreover, if $\mathcal{F}$ is a nonempty class of left R—modules closed under taking submodules, direct products, isomorphic copies, and injective hulls then there exists an injective left R—module E satisfying the condition that a left R—module A belongs to $\mathcal{F}$ if and only if it is E—torsionfree.

VERIFICATION: The first part of the observation is immediate. Now consider a nonempty class $\mathcal{F}$ of left R—modules closed under taking submodules, direct products, isomorphic copies, and injective hulls and let $\{Rm_i \mid i \in \Omega\}$ be a set of nonisomorphic cyclic left R—modules having the property that every cyclic left R—module in $\mathcal{F}$ is isomorphic to one of them. For each i in Ω, let H_i be an injective hull of Rm_i. Then $E = \Pi H_i$ is an injective left R—module that is a member of $\mathcal{F}$. Therefore any E—torsionfree left R—module is embeddable in a direct product of copies of E and also is a member of $\mathcal{F}$.

Conversely, if A is a member of $\mathcal{F}$ which has an injective hull C then C is also a member of $\mathcal{F}$. For each element c of C there is a direct summand H_c of C which is an injective hull of Rc. Let π_c be the projection of C onto H_c in this decomposition. Clearly $\cap\{\ker(\pi_c) \mid c \in C\} = (0)$ and so we have an R—monomorphism α from C to $\Pi\{H_c$

| c ∈ C} defined by $(x\alpha)(c) = x\pi_c$ for all x in C. This direct product is, in turn, embeddable in a direct product of copies of E. Therefore C (and hence A) is E–torsionfree.
□

Observation 1 characterizes classes of torsionfree modules in terms of certain closure properties. These are not the only closure properties that such classes have.

DEFINITION. A class of left R–modules is said to be *closed under extensions* if a left R–module A belongs to the class whenever there exists a submodule B of A such that both B and A/B belong to the class.

Verification of the following observation is left as an exercise.

OBSERVATION 2. If E is an injective left R–module then the class of E–torsionfree left R–modules is closed under extensions.

EXERCISES

1. Verify Observation 2.

2. Let R be an integral domain. Show that there exists an injective left R–module E satisfying the condition that a left R–module A is E–torsionfree if and only if ra ≠ 0 for every nonzero element r of R and every nonzero element a of A.

3. Let p be a prime element of a principal ideal domain R. Show that there exists an injective left R–module E satisfying the condition that a left R–module A is E–torsionfree if and only if $p^i a \neq 0$ for all nonnegative integers i and all nonzero elements a of A.

4. A left R—module A is *nonsingular* if for each nonzero element a of A and each large left ideal L of R there exists an element r of L satisfying ra $\neq$ 0. Show that there is an injective left R—module E satisfying the condition that a left R—module is E—torsionfree if and only if it is nonsingular.

§4 TORSION MODULES

If E is an injective left R—module then a left R—module A is *E—torsion* if and only if there are no nonzero homomorphisms from A to E. By Observation 1 of §2 we see that a left R—module A is E—torsion if and only if it is E'—torsion for any injective left R—module equivalent to E.

OBSERVATION 1. If E is an injective left R—module then the class of E—torsion left R—modules is closed under taking submodules, homomorphic images, coproducts, and extensions. Moreover, if $\mathcal{J}$ is a nonempty class of left R—modules closed under taking submodules, homomorphic images, coproducts, and extensions then there exists an injective left R—module E satisfying the condition that a left R—module A belongs to $\mathcal{J}$ if and only if it is E—torsion.

VERIFICATION: The closure of the class of E—torsion left R—modules under taking submodules follows from the injectivity of E: any nonzero homomorphism from a submodule of a left R—module A to E can be enlarged to a nonzero homomorphism from A to E. The closure of this class under taking homomorphic images, coproducts, and extensions is immediate. Now consider a nonempty class of left R—modules closed under taking submodules, homomorphic images, coproducts, and extensions. Let $\mathcal{B}$ be the class of all left R—modules B such that there is no nonzero map from any member of $\mathcal{J}$ to an injective hull of B. Let $\{Rm_i \mid i \in \Omega\}$ be a set of nonisomorphic cyclic left R–modules in $\mathcal{B}$ such that every cyclic left R–module in $\mathcal{B}$ is isomorphic to one of them. For each i in Ω, let H_i be an injective hull of Rm_i and let $E = \Pi\{H_i \mid i \in \Omega\}$. Then E is injective and, by construction, every module in $\mathcal{J}$ is E–torsion. We are left to show that every E–torsion left R–module belongs to $\mathcal{J}$.

Let B be an E–torsion left R–module and let U be the set of all elements b of B satisfying the condition that Rb belongs to $\mathcal{J}$. This set is nonempty since $0 \in U$. Moreover, if B' is the submodule of B generated by U then B' is a homomorphic image of $\amalg\{Rb \mid b \in U\}$ and so, by the closure conditions of $\mathcal{J}$, also belongs to $\mathcal{J}$. We are done if B' = B and so assume that this is not the case. If $x \in B \setminus B'$ and if $[Rx + B']/B'$ belongs to $\mathcal{J}$ then, by the closure of $\mathcal{J}$ under extensions, we conclude that $Rx + B'$ also belongs to $\mathcal{J}$, contradicting the choice of B'. Therefore no nonzero submodule of B/B' belongs to $\mathcal{J}$. Let C be an injective hull of B/B'. Since B/B' is large in C, it follows that no nonzero submodule of C can belong to $\mathcal{J}$. Since $\mathcal{J}$ is closed under taking homomorphic images, this implies that there is no nonzero map from any member of $\mathcal{J}$ to C. Thus B/B' belongs to $\mathcal{B}$. But this implies that there are no nonzero maps from B to C, which is impossible unless B = B'. Thus B belongs to $\mathcal{J}$. $\square$

The verification of Observation 1 also contains the clue to the relationship between the classes of E–torsion and of E–torsionfree modules. In fact, these classes can be used directly to define each other.

OBSERVATION 2. If E is an injective left R–module then:

(1) A left R–module A is E–torsion if and only if there is no nonzero map from A to an injective hull of any E–torsionfree left R–module B.

(2) A left R–module B is E–torsionfree if and only if

there is no nonzero map from any E–torsion left R–module A to any injective hull of B.

VERIFICATION: (1) Let A be an E–torsion left R–module and let B be an E–torsionfree left R–module having an injective hull C. Then there is an R–monomorphism λ from C to $E' = \Pi\{E_i \mid i \in \Omega\}$ for some set Ω, where each E_i is isomorphic to E. If there were a nonzero map $\alpha:A \to C$ then there would be an i in Ω such that $\alpha\lambda\pi_i$ would be nonzero, where $\pi_i:E' \to E_i$ is the projection onto the ith factor of E'. But this is impossible if A is E–torsion. Conversely, if there is no nonzero map from A to an injective hull of any E–torsionfree left R–module then, in particular, there is no such map from A to E, and so A is E–torsion.

(2) Let B be a left R–module having an injective hull C. If B is E–torsionfree and if A is a E–torsion left R–module then there is no nonzero map from A to C by (1). Conversely, assume that B satisfies the condition that there is no nonzero map from any E–torsion left R–module to C. Let H be the set of maps from C to E and, for each α in H, let E_α be a copy of E. Then we have a map β from C to $\Pi\{E_\alpha \mid \alpha \in H\}$ defined by $(c\beta)(\alpha) = c\alpha$ for all c in C and α in H. Let K be the kernel of β. Then any map from K to E is the restriction of one of the members of H and so is the zero map. Thus there are no nonzero maps from K to E and so K is E–torsion. But K is a submodule of C. By the assumption on B, this means that there is no nonzero map from K to C. Since K is a submodule of C, this means that K = (0). Therefore β is an R–monomorphism and so C is isomorphic to a submodule of a direct product of copies of E, proving that C, and hence B, is E–torsionfree. $\square$

In particular, we note that if A is an E–torsion left R–module and if B is an E–torsionfree left R–module then there are no nonzero maps from A to B. Also, as a consequence of the previous two observations, we see the following.

OBSERVATION 3. If E is an injective left R—module and if A is an arbitrary left R—module then there is a unique submodule A' of A satisfying: (1) A' is E—torsion; and (2) A/A' is E—torsionfree.

VERIFICATION: Take A' = ∩{ker(α) | α a map from A to E}. As in the verification of Observation 2, we see that there are no nonzero maps from A' to E (and so A' is E—torsion) while, on the other hand, A/A' is isomorphic to a submodule of a direct product of copies of E (and so is E—torsionfree). □

Note that the submodule A' of a left R—module A defined in Observation 3 must contain all E—torsion submodules of A for if A" is an E—torsion submodule of A not contained in A' then [A' + A"]/A' is a nonzero submodule of A/A' which is E—torsion since it is isomorphic to a homomorphic image of A". But E—torsionfree modules cannot have nonzero E—torsion submodules, so we have a contradiction. Thus we have verified the following observation.

OBSERVATION 4. If E is an injective left R—module and if A is an arbitrary left R—module, then A has a unique maximal E—torsion submodule.

We will denote the unique maximal E—torsion submodule of a left R—module A by $T_E(A)$. Note that a left R—module A is E—torsionfree precisely when $T_E(A) = (0)$.

The verification of the following observation is left as an exercise.

OBSERVATION 5. Let E be an injective left R—module and let B be a submodule of a left R—module A. If A' is the unique maximal E—torsion submodule of A, then the unique maximal E—torsion submodule of B is precisely A' ∩ B.

This observation is important in establishing the following result, which shows that if a module A having an injective hull C satisfies the condition that C/A is E–torsionfree, then A satisfies a weak injectivity condition. We will need this observation in Chapter 15.

OBSERVATION 6. Let A be a left R–module contained in an injective hull C and let E be an injective left R–module. Then C/A is E–torsionfree if and only if any map from a submodule B' of a left R–module B to A can be enlarged to a map from B to A whenever B/B' is E–torsion.

VERIFICATION: Suppose that C/A is E–torsionfree and let B' be a submodule of a left R–module B satisfying the condition that B/B' is E–torsion. If λ:A → C is the inclusion map and if α is a map from B' to A then $\lambda\alpha$ can be enlarged to a map β from B to C. Moreover, β induces a map β' from B/B' to C/A defined by $(b+B')\beta' = b\beta+A$ for all b in B. But B/B' is E–torsion and C/A is E–torsionfree so, by Observation 2, β' is the zero map. Therefore $\mathrm{im}(\beta) \subseteq A$.

Now, conversely, assume that the stated condition is satisfied, and let $C'/A = T_E(C/A)$. We are done if we can show that this is (0), namely that C' = A. By the given condition, the identity map ι:A → A can be enlarged to a map α from C' to A which must be a monomorphism since A is large in C' (for it is large in C) and $\ker(\alpha) \cap A = \ker(\iota) = (0)$. Moreover, $A = \mathrm{im}(\iota) \subseteq \mathrm{im}(\alpha) \subseteq A$, so α is an isomorphism. This can happen only if C' = A and $\beta = \iota$, which is all we needed to show. $\square$

Notice that we don't really need all submodules B' of B such that B/B' is E–torsion in Observation 6. A slight modification of the proof of Baer's criterion, which we leave as an exercise, will show that all that is necessary is to be able to enlarge all maps from H to A, where H is any left ideal of

R satisfying the condition that R/H is E-torsion.

EXERCISES

1. Let $\{E_i \mid i \in \Omega\}$ be a family of injective left R-modules and let $E = \Pi\{E_i \mid i \in \Omega\}$. Show that a left R-module is E-torsion if and only if it is E_i-torsion for each i in Ω.

2. Let H be a prime ideal of a commutative ring R and let E be an injective hull of R/H. Show that a left R-module A is E-torsion if and only if for each element a of A there exists an element r of $R \setminus H$ satisfying ra = 0.

3. Let R be an arbitrary ring. Show that there exists an injective left R-module E satisfying the condition that a left R-module A is E-torsion if and only if every factor module of A is projective.

4. Let R be a left noetherian ring. Show that there exists an injective left R-module E satisfying the condition that a left R-module A is E-torsion if and only if every factor module of an injective hull of A is injective.

5. Let E be a left R-module satisfying the condition that any nonzero E-torsionfree left R-module has a nonzero projective submodule. Show that the class of E-torsion left R-modules is closed under taking injective hulls.

6. Verify Observation 5.

7. Let E be an injective hull of R. If H is a left ideal of R satisfying the condition that R/H is E-torsion, show that H is large in R.

8. Let R be a commutative noetherian ring and let E be an injective left R-module. Show that the class of all E-torsion left R-modules is closed under taking injective hulls.

9. Let A be a left R—module contained in an injective hull C and let E be an injective left R—module. Show that C/A is E—torsionfree if and only if any map from a left ideal H of R to A can be enlarged to a map from R to A whenever R/H is E—torsion.

10. Show that injective left R—modules E and E' are equivalent if and only if a cyclic left R—module is E—torsion when and only when it is E'—torsion. Conclude from this that the family of all equivalence classes of injective left R—modules is in one—to—one correspondence with a set (of sets of left ideals of R).

11. Let R be a principal ideal domain. Show that there exists an injective left R—module E such that, for any left R—module A, the E—torsion submodule of A is $\{a \in A \mid \text{Ann}(a) \neq (0)\}$.

NOTES FOR CHAPTER 14

The notions of torsion and torsionfreeness at an injective left R—module were first developed systematically by Gabriel [1962] and Dickson [1966]. More detailed systematic treatments of this theory can be found in Lambek [1971] or in Golan [1975, 1986].

15

The Module of Quotients

§1 INTRODUCTION

In this chapter we move from the notions of modules torsion and torsionfree with respect to an equivalence class of injective (or of projective) left R—modules to the construction of modules of quotients at such an equivalence class. The construction is done in two stages: beginning with a left R—module A, we first move via the natural map to a torsionfree factor module of A; then we embed this factor module in a certain submodule of its injective hull which is determined by a weak form of injectivity. What is surprising, and not at all clear at this stage, is that such a construction is, to use a term from category theory, "functorial". Not only can we assign to each left R—module a module of quotients but also we can assign to each map between left R—modules a unique map between their modules of quotients.

The module of quotients thus constructed is hard to envision. It is a submodule of an injective hull and, since for the latter we have only an existence proof but not, in general, a method of construction, we do not know what it looks like. In

Section §4, however, we show that the module of quotients can always be considered as a submodule of a very specific (though extremely large) module of homomorphisms. This will be important in Chapter 15 for the representation of the ring of quotients in a specific form.

§2 THE MODULE OF QUOTIENTS DEFINED BY EQUIVALENT INJECTIVE MODULES

We are now going to construct the module of quotients of a left R–module A with respect to an equivalence class of injective left R–modules. Since this procedure is somewhat complicated, we will do it in stages. In defining our construction, we will use a particular injective left R–module E. However, note that at each stage the construction would not be altered if, at that stage, we replaced E by any other injective module equivalent to it, so that the module of quotients depends only on the equivalence class of E, and not on the particular module chosen from this class.

Let us begin therefore with an injective left R–module E and an arbitrary left R–module A.

(1) In Observations 3 and 4 of §14.4 we have seen that there exists a unique maximal E–torsion submodule $T_E(A)$ of A. Then $T_E(A)$ is E–torsion and $A/T_E(A)$ is E–torsionfree. For the purposes of this construction, denote the module $A/T_E(A)$ by $A_{(1)}$.

(2) Choose an injective hull C of $A_{(1)}$ and consider the factor module $A_{(2)} = C/A_{(1)}$.

(3) Now let $Q_E(A)/A_{(1)}$ equal $T_E(A_{(2)})$. It is this module $Q_E(A)$ which we define to be the *module of quotients* of A at the equivalence class of E.

Again, let us emphasize that $Q_E(A)$ depends not on E but only on the equivalence class of E. Also, the only choice we had in the above construction was the choice of the injective hull C. Since injective hulls of modules are unique up to isomorphism, this means that $Q_E(A)$ is defined uniquely up to isomorphism. We also note that the above construction also provided us with a map $\gamma_E(A)$ from A to $Q_E(A)$ which is the composite of the natural map $A \to A_{\{1\}}$ and the inclusion map of $A_{\{1\}}$ into $Q_E(A)$.

We now want to make a series of observations concerning some striking properties of the module $Q_E(A)$ which we have constructed. The first is an immediate consequence of the construction of $Q_E(A)$.

OBSERVATION 1. If E is an injective left R–module and if A is an arbitrary left R-module then:
(1) $\ker(\gamma_E(A))$ is E–torsion;
(2) $Q_E(A)/\mathrm{im}(\gamma_E(A))$ is E–torsion;
(3) $Q_E(A)$ is E–torsionfree; and
(4) If C is an injective hull of $Q_E(A)$ then $C/Q_E(A)$ is E–torsionfree.

The second observation shows a crucial property of $Q_E(A)$.

OBSERVATION 2. Let B' be a submodule of a left R–module B satisfying the condition that B/B' is E–torsion. Then any map from B' to $Q_E(A)$ can be uniquely enlarged to a map from B to $Q_E(A)$.

VERIFICATION: Let α be a map from B' to $Q_E(A)$ and let λ be the inclusion map $Q_E(A) \to C$. By (4) of Observation 1 and by Observation 6 of §14.4, α can be enlarged to a map β from B to $Q_E(A)$. We are left to show that β is unique. Assume that β_1 and β_2 are maps from B to $Q_E(A)$ both of which enlarge α. Then there is a map δ from B to $Q_E(A)$ defined by $b\delta = b\beta_1 - b\beta_2$, the kernel of

which contains B'. Therefore δ defines a map δ' from
B/B' to $Q_E(A)$ by $(b+B')\delta' = b\delta$ for all b in B. But, by
hypothesis, B/B' is E–torsion while, by Observation 1, $Q_E(A)$
is E–torsionfree. Therefore, by Observation 2 of §14.4, δ' is
the zero map, proving that $\beta_1 = \beta_2$. $\square$

The next observation shows that the properties listed in
Observation 1 essentially characterize the module of quotients.

OBSERVATION 3. If $\alpha:A \to B$ is a map of left
R–modules satisfying the following conditions:
(1) ker(α) is E–torsion;
(2) B/im(α) is E–torsion;
(3) B is E–torsionfree; and
(4) If D is an injective hull of B then D/B is
    E–torsionfree then there exists an isomorphism $\beta:B \to$
    $Q_E(A)$ satisfying $\gamma_E(A) = \alpha\beta$.

VERIFICATION: Let $T_E(A)$ be the unique maximal
E–torsion submodule of A. Then, by (1), ker(α) $\subseteq T_E(A)$. On
the other hand, $T_E(A)\alpha$ is an E–torsion submodule of B.
Since B is E–torsionfree, this means that $T_E(A)\alpha = (0)$ and
so $T_E(A) \subseteq$ ker(α), establishing equality. Therefore, since
$T_E(A)$ is also precisely the kernel of $\gamma_E(A)$, it suffices for the
rest of the proof to replace A by $A/T_E(A)$ and so assume
that A is E–torsionfree and that the maps α and $\gamma_E(A)$
are monomorphisms. Since B/im(α) is E–torsion, we know by
Observation 2 that $\gamma_E(A)$ can be enlarged uniquely to a map
β from B to $Q_E(A)$ which satisfies $\gamma_E(A) = \alpha\beta$. On the
other hand, $Q_E(A/im(\gamma_E(A)))$ is E–torsion and the same
argument used in verifying Observation 2, together with (3) and
(4), shows that $\gamma_E(A)$ can be enlarged uniquely to a map
$\beta':Q_E(A) \to B$ which satisfies $\alpha = \gamma_E(A)\beta'$. Then $\beta\beta'$ must
be the unique endomorphism θ of B satisfying $\alpha\theta = \alpha$
and $\beta'\beta$ must be the unique endomorphism ψ of $Q_E(A)$

satisfying $\gamma_E(A) = \gamma_E(A)\psi$. This means that $\beta\beta'$ is the identity map on B and $\beta'\beta$ is the identity map on $Q_E(A)$, which suffies to prove that β is an isomorphism and that $\beta' = \beta^{-1}$. $\square$

We want to be sure, of course, that this construction really generalizes the construction of $S^{-1}A$, as defined in §14.1, in the case that S is a multiplicatively–closed subset of a commutative ring R which does not contain 0. So consider such a subset S and let A be a left R–module. Then we have a map $\theta:A \rightarrow S^{-1}A$ which sends an element a of A to sa/s of $S^{-1}A$ (here s is any element of S; they all yield the same element of $S^{-1}A$). By Observation 1 of §14.4 we observe that there is an injective left R–module E satisfying the condition that a left R–module B is E–torsion if and only if for every element b of B there exists an element s of S satisfying $sb = 0$. In particular, if $a \in \ker(\theta)$ then $sa/s = 0$ so there exists an element s'' of S satisfying $(s''s)a = s''(sa) = 0$ and so $\ker(\theta)$ is E–torsion. If a/s is an arbitrary element of $S^{-1}A$ then $sa/s \in \text{im}(\theta)$ and so $s(a/s) \in \text{im}(\theta)$. This implies that $S^{-1}A/\text{im}(\theta)$ is also E–torsion. On the other hand, assume that $a/s \in T_E(S^{-1}A)$. Then there exists an element s' of S satisfying $s'a/s = 0$. But then $a/s = 0/s' = 0$ and so $a/s = 0$. Thus $S^{-1}A$ itself is E–torsionfree. Finally, let $S^{-1}A$ be contained in an injective hull C. To show that $C/S^{-1}A$ is E–torsionfree it suffices, as we observed at the end of Section §14.4, to show that if H is a left ideal of R satisfying the condition that R/H is E–torsion then any map from H to $S^{-1}A$ can be enlarged to a map from R to $S^{-1}A$. Let $\alpha:H \rightarrow S^{-1}A$ be such a map. Since R/H is E–torsion, there exists an element s of S satisfying $s(1+H) = 0+H$ and so $s \in S \cap H$. Then $s\alpha = a/t = s(a/st)$ for some elements a of A and t of S. If h is an arbitrary element of H and if $h\alpha = a'/s'$ for some a' in A and s' in S then, by commutativity, $sa'/s' = s(h\alpha) = (sh)\alpha = (hs)\alpha = h(s\alpha) = hs(a/ts) = s(ha/ts)$ and so $a'/s' = ha/ts$ in $S^{-1}A$.

This shows that $h\alpha = h(a/ts)$ for all h in H and so the map from R to $S^{-1}A$ which sends an element r of R to ra/ts enlarges α. We are now in a position to apply Observation 3 and to conclude with the following observation.

> **OBSERVATION** 4. Let S be a multiplicatively-closed subset of a commutative ring R which does not contain 0 and let E be an injective left R-module satisfying the condition that a left R-module B is E-torsion if and only if for each element b of B there exists an element s of S satisfying $sb = 0$. Then $S^{-1}A \cong Q_E(A)$ for any left R-module A.

§3   THE MODULE OF QUOTIENTS DEFINED BY EQUIVALENT PROJECTIVE MODULES

In the previous section we defined the module of quotients of an arbitrary left R-module given by an equivalence class of injective left R-modules. Can we carry out a similar construction for projective left R-modules? First we need an appropriate definition of equivalence.

> **DEFINITION.** Projective left R-modules P and P' are *equivalent* if each of them is isomorphic to a homomorphic image (and hence a direct summand) of a coproduct of copies of the other.

If this condition looks somewhat familiar, it is because we have already encountered it in Chapter 5. In fact, we can now restate Observation 4 of §5.2 as follows:

> **OBSERVATION** 1. Projective left R-modules P and P' are equivalent if and only if they have the same trace.

If P is a projective left R–module and A is an arbitrary left R–module, we will say that A is *P-torsion* if and only if ra = 0 for all a in A and all r in tr(P). Observation 1 then assures us that A is P–torsion if and only if it is P'–torsion for any projective left R–module P' equivalent to P. We then notice the following:

OBSERVATION 2. Let P be a projective left R–module. Then the class of all P–torsion left R–modules is closed under taking submodules, homomorphic images, direct products, and extensions.

VERIFICATION: The closure of the class of all P–torsion left R–modules under taking submodules, homomorphic images, and direct products is an immediate consequence of the definition. Now suppose that B is a submodule of a left R–module A satisfying the condition that both B and A/B are P–torsion. If a is an element of B then we know that ra = 0 for all r in tr(P). If $a \in A \setminus B$ and $r \in$ tr(P) then, by Observation 3 of §5.2 we see that $r = \Sigma s_i t_i$, where the s_i and t_i are elements of tr(P). Therefore $ra = \Sigma s_i t_i a$. But $t_i a \in B$ for all i since A/B is P–torsion and so $s_i(t_i a) = 0$ for all i. Therefore ra = 0, proving that A is P–torsion. □

Since the coproduct of a family of left R–modules is a submodule of their direct product, we see from Observation 2 that if P is a projective left R–module then the family of all P–torsion left R–modules is also closed under taking coproducts. We can therefore apply Observation 1 of §14.3 and note that there exists an injective left R–module E(P), determined up to equivalence of injective modules, which satisfies the condition that a left R–module is E(P)–torsion if and only if it is P–torsion. We say that a left R–module is *P-torsionfree* when it is E(P)–torsionfree. For any left R–module A, we can construct the module of quotients of A at E(P), which we will denote by $Q_P(A)$, and call the *module of quotients* of

A at the equivalence class of the projective module P. As it turns out, this module can be described in a rather nice way.

OBSERVATION 3. Let P be a projective left R–module and let A be an arbitrary left R–module. Then $Q_P(A)$ is isomorphic to $\text{Hom}_R(\text{tr}(P), A/T_P(A))$, where $T_P(A)$ is the unique maximal P–torsion submodule of A.

VERIFICATION: Let $B = \text{Hom}_R(\text{tr}(P), A/T_P(A))$. We note that the left R–module $R/\text{tr}(P)$ is P–torsion and so, by Observation 2 of §15.2, any map α in B can be enlarged to a unique map α^* from R to $Q_P(A)$. The function β from B to $Q_P(A)$ which sends a map α to $1\alpha^*$ can be easily verified to be a map of left R–modules, which is well–defined by the uniqueness of α^*. If $1\alpha_1{}^* = 1\alpha_2{}^*$ for elements α_1 and α_2 of B then $\alpha_1 - \alpha_2$ induces a map from $R/\text{tr}(P)$ to $A/T_P(A)$, which must be the zero map since $R/\text{tr}(P)$ is P–torsion and $A/T_P(A)$ is P–torsionfree. Therefore $\alpha_1 = \alpha_2$, and we have shown that the function β is one–to–one. If x $\in$ $Q_P(A)$ then x determines a map δ from R to $Q_P(A)$ given by $r\delta = rx$. Let $\nu : Q_P(A) \to Q_P(A)/[A/T_P(A)]$ be the natural map. Then $\text{im}(\delta\nu)$ is P–torsion, and thus so is $R/\text{ker}(\delta\nu)$. This implies that $\text{tr}(P)[R/\text{ker}(\delta\nu)] = (0)$ and so $\text{tr}(P) \subseteq \text{ker}(\delta\nu)$. If α is the restriction of δ to $\text{tr}(P)$, we then see that $\alpha \in B$ and, by uniqueness, that $\alpha^* = \delta$. Therefore $x = 1\delta = 1\alpha^*$, showing that the map β is an R–epimorphism and hence an isomorphism. $\square$

EXERCISES

1. Let E be an injective left R–module satisfying the condition that the class of all E–torsion left R–modules is closed under taking direct products. Show that there exists a projective left R–module P such that a left R–module is E–torsion if and only if it is P–torsion.

2. For an injective left R−module E, show that the following conditions are equivalent:
(a) There exists a projective left R−module P such that a left R−module is E−torsion if and only if it is P−torsion.
(b) The set of all left ideals L of R satisfying the condition that R/L is E−torsion has a unique minimal element.

3. Let R be a commutative noetherian ring.
(a) If I is a finitely−generated ideal of R satisfying $I^2 = I$, show that I = Re for some idempotent element e of R.
(b) Let P be a projective left R−module. For any left R−module A, show that the unique maximal P−torsion submodule of A is a direct summand of A.

§4 A REPRESENTATION OF THE MODULE OF QUOTIENTS

Let E be an injective left R−module having an endomorphism ring S and let A be an arbitrary left R−module. As we saw in §0.5, both E and $Hom_R(A,E)$ are right S−modules. Similar reasoning leads us to see that $Hom_S(Hom_R(A,E),E)$ is a left R−module which, for simplicity, we will denote by $B_E(A)$. For each left R−module A there is a standard map $\delta_E(A)$ from A to $B_E(A)$ defined in the following manner: if a is an element of A then $a\delta_E(A)$ is the evaluation S−homomorphism which sends any element α of $Hom_R(A,E)$ to $a\alpha$.

OBSERVATION 1. If E is an injective left R−module and if A is an arbitrary left R−module then the kernel of $\delta_E(A){:}A \to B_E(A)$ is $T_E(A)$.

VERIFICATION: If $a \in T_E(A)$ then Ra is E−torsion. Since E is E−torsionfree, this means that $a\alpha = 0$ for any α

in $\mathrm{Hom_R(A,E)}$ and so $a \in \ker(\delta_E(A))$. Thus $T_E(A) \subseteq \ker(\delta_E(A))$. Conversely, assume that α is a map from $\ker(\delta_E(A))$ to E. Then α can be enlarged to a map β from A to E. If α is not the zero map then there exists an element x of $\ker(\delta_E(A))$ satisfying $x\alpha \neq 0$ and so $x\beta \neq 0$. But $x\beta = (x\delta_E(A))(\beta)$, contradicting the assumption that x belongs to $\ker(\delta_E(A))$. Therefore there are no nonzero maps from $\ker(\delta_E(A))$ to E and so, by Observation 2 of §14.4, $\ker(\delta_E(A))$ is E-torsion and so is contained in $T_E(A)$. Thus $\ker(\delta_E(A)) = T_E(A)$. $\square$

This observation leads us to the following very important conclusion.

> **OBSERVATION** 2. If E is an injective left R-module and if A is an arbitrary left R-module then there is a unique R-monomorphism $\varepsilon_E(A)$ from $Q_E(A)$ to $B_E(A)$ satisfying $\gamma_E(A)\varepsilon_E(A) = \delta_E(A)$.

VERIFICATION: Let S be the endomorphism ring of E. Then $N = \mathrm{Hom_R(A,E)}$ is a right S-module. We know that there exists a free right S-module Y_1 and an S-epimorphism φ_1 from Y_1 to N. Moreover, there exists another free right S-module Y_2 and an S-epimorphism φ_2 from Y_2 to $\ker(\varphi_1)$. The map φ_1 induces a homomorphism of left R-modules $\alpha_1{:}\mathrm{Hom_S(N,E)} \rightarrow \mathrm{Hom_S(Y_1,E)}$ which sends an S-homomorphism δ to $\delta\varphi_1$. (Remember that these functions are written as acting on the *left* since we are talking about *right* S-modules!) Since φ_1 is an epimorphism, α_1 is a monomorphism because $\delta\varphi_1$ can be the zero-map only when δ is. Similarly, φ_2 induces a homomorphism of left R-modules $\alpha_2{:}\mathrm{Hom_S(Y_1,E)} \rightarrow \mathrm{Hom_S(Y_2,E)}$. Since the image of φ_2 equals the kernel of φ_1, it is not difficult to check (and we expect you to do so) that the kernel of α_2 is precisely the image of α_1.

The free right S-modules Y_1 and Y_2 are isomorphic to coproducts of copies of S and so, by using the right-left inversion of Observation 2 of §0.6, we see that $\mathrm{Hom}_S(Y_1, N)$ and $\mathrm{Hom}_S(Y_2, N)$ are isomorphic to direct products of copies of E. In summary, what we have is the following:

(1) $B_E(A) = \mathrm{Hom}_S(N, E)$ is isomorphic to a submodule C of a direct product D of copies of E.

(2) D/C is also isomorphic to a submodule of a direct product D' of copies of E.

The first of these conclusions simply says that $\mathrm{Hom}_S(N, E)$ is E–torsionfree. Since D is injective, it contains a submodule C' containing C which is an injective hull of C. Therefore C'/C is isomorphic to a submodule of D' and hence is E–torsionfree. By Observation 6 of §14.4, this means that if B' is a submodule of a left R–module B satisfying the condition that B/B' is E–torsion then any map α from B' to $B_E(A)$ can be enlarged to a map from B to $B_E(A)$. If β_1 and β_2 are two such enlargements then $\beta_1 - \beta_2$ induces a map from B/B' to $B_E(A)$, which must be the zero map since B/B' is E–torsion and $B_E(A)$ is E–torsionfree. Therefore such enlargements are unique.

Since $\ker(\gamma_E(A)) = T_E(A) = \ker(\delta_E(A))$, we see that $\gamma_E(A)$ induces a monomorphism from $A/T_E(A)$ to $Q_E(A)$ with image A' such that $Q_E(A)/A'$ is E–torsion. Also, the map $\delta_E(A)$ induces an R–monomorphism from $A/T_E(A)$ to $B_E(A)$ which, by the above, can be uniquely enlarged to a map $\varepsilon_E(A)$ from $Q_E(A)$ to $B_E(A)$. Moreover, $\varepsilon_E(A)$ is a monomorphism since A' is large in $Q_E(A)$. By construction, $\gamma_E(A)\varepsilon_E(A) = \delta_E(A)$. $\square$

Thus we have seen that $Q_E(A)$ is always isomorphic to a submodule of $B_E(A)$. These modules are not equal in general, but they are in one very important case, the verification of which we leave as an exercise.

OBSERVATION 3. Let E be an injective left R–module and let A be a left R–module satisfying the condition that $Q_E(A)$ is isomorphic to a subset of a direct product of finitely—many of copies of E. Then $\varepsilon_E(A)$ is an isomorphism.

EXERCISE

1. Verify Observation 3 by checking through the proof of Observation 2 very carefully for such left R—modules A.

NOTES FOR CHAPTER 15

The material in this chapter is all based on Gabriel [1962], Dickson [1966], and Lambek [1971]. For general expositions of the theory, see Golan [1975, 1986].

16

Rings of Quotients

§1   INTRODUCTION

We are almost finished. Having shown how to construct the module of quotients $Q_E(A)$ of any left R–module A at an equivalence class of injective modules, we are left only to show that $Q_E(R)$ is in fact a ring and that $Q_E(A)$ is, in a natural way, a left $Q_E(R)$–module for every left R–module A. We do this by a trick: namely we show that $Q_E(R)$ is isomorphic, as a left R–module, to its own endomorphism ring R_E. Finally, we use the representation theory of the quotient module developed in Section §15.4 to give a characterization of this ring, namely we show that any injective left R–module E is equivalent to an injective left R–module E' having endomorphism ring S such that R_E is isomorphic, as a ring, to $Hom_S(E',E')$.

§2.   RINGS OF QUOTIENTS

If E is an injective left R–module then, for any left R–module A, we have already seen how to construct the module of quotients $Q_E(A)$ at the equivalence class of E. In particular we can do this for the ring R considered as a left module over itself. Then $\text{Hom}_R(Q_E(R),Q_E(R))$ is a ring, which we will call the *ring of quotients* of R at the equivalence class of E. We will denote this ring simply by R_E.

OBSERVATION 1. For any injective left R–module E the ring R_E has the structure of a left R–module isomorphic to $Q_E(R)$.

VERIFICATION: An element q of $Q_E(R)$ induces a map $\beta_q : R \to Q_E(R)$ which sends an element r of R to rq. Since $Q_E(R)$ is E–torsionfree we know that $T_E(R) \subseteq \ker(\beta_q)$ for any such q. Therefore each β_q induces a map from $R/T_E(R)$ to $Q_E(R)$ which, as a consequence of Observation 2 of §15.2, can be uniquely extended to an R–endomorphism θ_q of $Q_E(R)$.

If r is an element of R we will denote the element r + $T_E(R)$ of $R/T_E(R)$ by r*. Then we can define the structure of a left R–module on R_E by setting $r\alpha = \theta_{r*}\alpha$. (The proof that this does in fact turn R_E into a left R–module is straightforward using the uniqueness of θ_q; we will leave it to you to complete.) Let $\theta : Q_E(R) \to R_E$ be the function which sends an element q of $Q_E(R)$ to θ_q. We claim that this is an R–homomorphism. Indeed, if x, y $\in Q_E(R)$ then $x\theta + y\theta$ and $(x + y)\theta$ both extend the map from $R/T_E(R)$ to $Q_E(R)$ induced by $\beta_x + \beta_y$ and so must be equal. Similarly, if r $\in$ R and x $\in Q_E(R)$ then $r(x\theta)$ and $(rx)\theta$ are equal. This establishes the claim. We note, in fact, that θ is an R–epimorphism for if $\alpha \in R_E$ then α is the image of 1*α under θ. Finally, θ is also a monomorphism since x $\neq$ 0 implies that $1*\theta_x = x \neq 0$ and so x $\neq 0$. Thus θ is the isomorphism we want. $\square$

OBSERVATION 2. If E is an injective left R–module and if A is an arbitrary left R–module then $Q_E(A)$ is, in a natural way, a left R_E–module.

VERIFICATION: If x is an element of $Q_E(A)$ then x defines an R– homomorphism from R to $Q_E(A)$ which sends an element r of R to rx. Then $T_E(R)$ is contained in the kernel of this map since $Q_E(A)$ is E–torsionfree and so we have an induced R–homomorphism from $R/T_E(R)$ to $Q_E(A)$ which, by Observation 2 of §15.2, is uniquely extendable to a map β_x from $Q_E(R)$ to $Q_E(A)$. If $\alpha \in R_E$, define $\alpha \cdot x$ to be $(1)\alpha\beta_x \in Q_E(A)$. A straightforward verification shows that this defines the structure of a left R_E–module on $Q_E(A)$. $\square$

We will end by constructing a more interesting representation of the ring of quotients of the ring R at an equivalence class of injective left R–modules. Let E be an injective left R–module having endomorphism ring S and let T = $\text{Hom}_S(E,E)$. This ring is called the *biendomorphism ring* of E. Then E is a left T–module. Moreover, there exists a ring homomorphism $\rho: R \rightarrow T$ defined as follows: if $r \in R$ then $\rho(r)$ is the S–endomorphism of E which sends an element x to rx. This ring homomorphism also induces the structure of a left R–module on T: if $\sigma \in T$ and $r \in R$, we set $r \cdot \sigma = \rho(r)\sigma$.

Now suppose that E is a cyclic right S–module, say E = xS. Then we have an R–homomorphism from T to E which sends an element σ of T to σx. This map is a monomorphism since $\sigma x = 0$ implies that $sy = 0$ for all y in E and so σ is the zero map. Also, we know that the right S–modules E and $\text{Hom}_R(R,E)$ are isomorphic and so T is isomorphic, as a left R–module, to $\text{Hom}_S(\text{Hom}_R(R,E),E)$ = $B_E(R)$. Thus we see that $B_E(R)$ is isomorphic to a submodule of E. By Observation 2 of §15.4, we see that $Q_E(R)$ is isomorphic to a submodule of E and the, by Observation 3 of §15.4, we must have an isomorphism between

$Q_E(R)$ and T. Therefore R_E and T are isomorphic as left R–modules. We leave it as an exercise to show that they are isomorphic as rings.

We now recall two very important points: (1) All of our constructions do not depend on the particular injective module chosen but only on the equivalence class of this module, and (2) by Observation 2 of §14.2, every injective left R–module is equivalent to an injective module which is cyclic as a right module over its endomorphism ring. Putting all of these pieces together, we conclude with the following characterization of quotient rings.

> **THEOREM** 12. If E is an injective left R–module then there is an injective left R–module E' equivalent to E such that R_E is isomorphic to the biendomorphism ring of E'.

Let us end with an example to illustrate this result. If R = $\mathbb{Z}$ and E = $\mathbb{Q}$ then $R_E = \mathbb{Q}$. Moreover, by Exercise 10 of §0.4, we know that the endomorphism ring of the left $\mathbb{Z}$–module $\mathbb{Q}$ is isomorphic to $\mathbb{Q}$. Therefore the biendomorphism ring of $\mathbb{Q}$ is also isomorphic to $\mathbb{Q}$. Thus we see that $\mathbb{Q}$ is the ring of quotients of $\mathbb{Z}$ at the injective left $\mathbb{Z}$–module $\mathbb{Q}$, which is what we certainly would expect.

EXERCISES

1. Let E be an injective left R–module and let A and B be arbitrary left R–modules. Show that any R–homomorphism from $Q_E(A)$ to $Q_E(B)$ is also an R_E–homomorphism.

2. Let E be an injective left R–module and let A be a left R–module. Show that $Q_E(A)$ is injective as a left R_E–module if and only if it is injective as a left R–module.

3. Let E be an injective left R–module having endomorphism

ring S and let $T = \text{Hom}_S(E,E)$.

(a) Show that the ring homomorphism from R to T induces a ring isomorphism between $R/T_E(R)$ and a subring T' of T.

(b) If E is cyclic as a right S-module, show that R_E and T are isomorphic by showing that the R-module isomorphism between them constructed above extends the ring isomorphism between the subring of R_E isomorphic to $R/T_E(R)$ and T'. (Use repeatedly the uniqueness of extensions of maps, as was done in the verification of Observation 1.)

NOTES FOR CHAPTER 16

The material in this chapter is based primarily on the work of Lambek [1971, 1972] and Morita [1970]. Other expositions can be found in Golan [1975, 1986].

References

F. Albrecht,   On projective modules over semi–hereditary rings, *Proceedings of the American Mathematical Society* 12 (1961), 638–639.

F. W. Anderson & K. R. Fuller, *Rings and categories of modules*, Berlin: Springer–Verlag, 1974.

E. Artin,   The influence of J. H. M. Wedderburn on the development of modern algebra, *Bulletin of the American Mathematical Society* 56 (1950), 65–72.

M. Auslander & D. Buchsbaum, *Groups, rings, modules*, New York: Harper & Row, 1974.

R. Baer,   Abelian groups that are direct summands of every containing abelian group, *Bulletin of the American Mathematical Society* 46 (1940), 800–806.

H. Bass,   Finitistic dimension and a homological generalization of semi–primary rings, *Transactions of the American Mathematical Society* 95 (1960), 466–488.

—————, *Algebraic K–theory*, New York: Benjamin, 1968.

I. Beck, Σ–injective modules, *Journal of Algebra* 21 (1972), 232–249.

L. Bican, P. Jambor, T. Kepka & P. Nemec,   A note on test modules, *Commentationes Mathematicae Universitatis Carolinae* 17 (1976), 345–355.

A. Blass,   Injectivity, projectivity, and the axiom of choice,

Transactions of the American Mathematical Society 225 (1979), 31–59.

T. S. Blyth, *Module theory*, Oxford: Oxford University Press, 1977.

R. T. Bumby, Modules which are isomorphic to submodules of each other, *Archiv der Mathematik* 16 (1965), 184–185.

M. Burrow, *Representation theory of finite groups*, New York: Academic Press, 1965.

E. Cartan & S. Eilenberg, *Homological algebra*, Princeton, N.J.: Princeton University Press, 1956.

B.-S. Chwe & J. Neggers, On the extension of linearly independent subsets of free modules to bases, *Proceedings of the American Mathematical Society* 24 (1970), 466–470.

P. M. Cohn, Some remarks on the invariant basis property, *Topology* 5 (1966), 215–228.

––––––, *Free rings and their relations* (2nd ed.), London: Academic Press, 1971.

C. W. Curtis & I. Reiner, *Representation theory of finite groups and associative algebras*, New York: Wiley–Interscience, 1962.

J. Dauns, *A concrete approach to division rings*, Berlin: Heldermann–Verlag, 1982.

S. Dickson, A torsion theory for abelian categories, *Transactions of the American Mathematical Society* 121 (1966), 223–235.

B. Eckmann & O. Schopf, Über injektiv Moduln, *Archiv der Mathematik* 4 (1953), 75–78.

C. Everett, Jr., Vector spaces over rings, *Bulletin of the American Mathematical Society* 48 (1942), 312–316.

C. Faith, Rings with ascending chain condition on annihilators, *Nagoya Mathematical Journal* 27 (1966), 179–191.

––––––, *Lectures on injective modules and quotient rings*, Lecture Notes in Mathematics #49, Berlin: Springer–Verlag, 1967.

––––––, *Algebra: Rings, modules and categories I*, Berlin:

Springer–Verlag, 1972.

–––––, *Algebra II: Ring theory*, Berlin: Springer–Verlag, 1976.

––––– & Y. Utumi, Baer modules, *Archiv der Mathematik* 15 (1964), 266–270.

––––– & E. Walker, Direct–sum representations of injective modules, *Journal of Algebra* 5 (1967), 203–221.

W. Feit, *The representation theory of finite groups*, Amsterdam: North Holland, 1982.

L. Fuchs, *Infinite abelian groups* (2 vols.), New York: Academic Press, 1970–3.

P. Gabriel, Des catégories abéliennes, *Bulletin de la Société Mathématique de France* 90 (1962), 323–448.

J. S. Golan, *Localization of noncommutative rings*, New York: Marcel Dekker, 1975.

–––––, *Torsion theories*, Harlow: Longman Scientific & Technical, 1986.

K. R. Goodearl, *Ring theory: nonsingular rings and modules*, New York: Marcel Dekker, 1976.

–––––, *Von Neumann regular rings*, London: Pitman, 1979.

M. Gray, *A radical approach to algebra*, Reading, Mass.: Addison–Wesley, 1970.

P. R. Halmos, *Naive set theory*, New York: Van Nostrand, 1960.

B. Hartley & T. O. Hawkes, *Rings, modules, and linear algebra*, London: Chapman and Hall Ltd., 1970.

T. Head, *Modules: a primer of structure theorems*, Monterey, Calif.: Brooks/Cole, 1974.

O. Helmer, Divisibility properties of integral functions, *Duke Mathematical Journal* 6 (1940), 345–356.

I. N. Herstein, *Topics in algebra*, Waltham Mass.: Blaisdell, 1964.

P. Hill, On the decomposition of certain infinite nilpotent groups, *Mathematische Zeitschrifte* 113 (1970), 237–248.

P. Hilton, *Lectures in homological algebra*, Regional Conference Series in Mathematics #8, Providence, R.I.: American Mathematical Society, 1971.

–––––– & Y.-C. Wu, *A course in modern algebra*, New York: Wiley–Interscience, 1974.

W. Hodges, Krull implies Zorn, *Journal of the London Mathematical Society* (2) 19 (1979), 285–287.

S. T. Hu, *Elements of modern algebra*, San Francisco: Holden–Day, 1965.

–––––, *Introduction to homological algebra*, San Francisco: Holden–Day, 1968.

J. Irwin & F. Richman, Direct sums of countable groups and related concepts, *Journal of Algebra* 2 (1965), 443–450.

N. Jacobson, *Basic Algebra I*, San Francisco: Freeman, 1974.

–––––, *Basic Algebra II*, San Francisco: Freeman, 1980.

J. P. Jans, Projective injective modules, *Pacific Journal of Mathematics* 9 (1959), 1103–1108.

I. Kaplansky, Projective modules, *Annals of Mathematics* 68 (1958), 372–377.

–––––, *Infinite abelian groups* (2nd ed), Ann Arbor: University of Michigan Press, 1969.

–––––, *Commutative rings* (rev. ed.), Chicago: University of Chicago Press, 1974.

F. Kasch, *Moduln und Ringe*, Stuttgart: B. G. Teubner, 1977.

J. L. Kelley, *General topology*, New York: Van Nostrand, 1955.

T. Y. Lam, *Serre's conjecture*, Lecture Notes in Mathematics #635, Berlin: Springer–Verlag, 1978.

J. Lambek, *Lectures on rings and modules*, Waltham, Mass: Blaisdell, 1966.

–––––, *Torsion theories, additive semantics, and rings of quotients*, Lecture Notes in Mathematics #177. Berlin: Springer–Verlag, 1971.

–––––, Bicommutators of nice injectives, *Journal of Algebra* 21 (1972), 60–73.

W. Leavitt, Modules over commutative rings, *American Mathematical Monthly* 71 (1964), 1112–1113.

H. Lenzing, A homological characterization of Steinitz rings, *Proceedings of the American Mathematical Society* 29 (1971), 269–271.

S. Mac Lane, *Categories for the working mathematician*, Berlin: Springer–Verlag, 1971.

————— & G. Birkhoff, *Algebra*, New York: Macmillan, 1967.

M. Marcus, *Introduction to modern algebra*, New York: Marcel Dekker, 1978.

H. Maschke, Über den arithmetischen Charakter der Coefficienten der Substitutionen endlicher linearer Substitutionsgruppen, *Mathematische Annalen* 50 (1898), 482–498.

E. Matlis, Injective modules over noetherian rings, *Pacific Journal of Mathematics* 8 (1958), 511–528.

N. McCoy, *The theory of rings*, New York: Macmillan, 1964.

B. R. McDonald, *Linear algebra over commutative rings*, New York: Marcel Dekker, 1984.

O. R. Mitchell & R. W. Mitchell, *An introduction to abstract algebra*, Monterey, Calif.: Brooks/Cole, 1970.

K. Morita, Localizations in categories of modules. I, *Mathematische Zeitschrift* 114 (1970), 121–144.

T. Nakayama, On Frobeniusean algebras I, *Annals of Mathematics* 40 (1939), 611–633.

M. Ojanguren & R. Sridharan, Cancellation of Azumaya algebras, *Journal of Algebra* 18 (1971), 501–505.

K. Oshiro, Lifting modules, extending modules, and their applications to QF–rings, *Hokkaido Mathematical Journal* 13 (1984), 310–338.

B. Osofsky, A generalization of quasi–Frobenius rings, *Journal of Algebra* 4 (1966), 373–387.

—————, Minimal cogenerators need not be unique, preprint, 1990.

D. Passman, *The algebraic structure of group rings*, New York: Wiley–Interscience, 1977.

R. Peinado, Note on modules, *Mathematics Magazine* 37 (1964), 266–267.

R. Rentschler, Sur les modules M tels que Hom(M,_) commute avec les sommes directes, *Comptes Rendus Hebdominaires de l'Académie de Science, Paris, Serie A*, 268 (1969), 930–933.

F. Richman, *Number theory: an introduction to algebra*, Monterey, Calif.: Brooks/Cole, 1971.

J. Rotman, *Notes on homological algebra*, New York: Van Nostrand Reinhold, 1970.

—————, *An introduction to homological algebra*, New York: Academic Press, 1979.

—————, *An introduction to the theory of groups* (3rd ed.), Boston: Allyn and Bacon, 1984.

B. Roux, Modules générateurs et cogénérateurs relatifs, *Bulletin des Sciences Mathématiques*, 2ᵉ Sèrie, 96 (1972), 97–110.

H. Rubin & J. E. Rubin, *Equivalents of the axiom of choice*, Amsterdam: North Holland, 1963.

—————, *Equivalents of the axiom of choice, II*, Amsterdam: North Holland, 1985.

W. Rudin, *Real and complex analysis*, New York: McGraw–Hill, 1966.

E. Rutter Jr., Two characterizations of QF rings, *Pacific Journal of Mathematics* 30 (1969), 777–784.

—————, PF modules, *Tohoku Mathematical Journal* 23 (1971), 201–206.

—————, A characterization of QF-3 rings, *Pacific Journal of Mathematics* 51 (1974), 533–536.

O. Schreier & E. Sperner, *Introduction to modern algebra and matrix theory*, New York: Chelsea, 1959.

D. W. Sharpe & P. Vámos, *Injective modules*, Cambridge Tracts in Mathematics #62, Cambridge: Cambridge University Press, 1972.

S. Shelah, Infinite abelian groups, Whitehead problem, and some constructions, *Israel Journal of Mathematics* 18 (1974), 243–256.

P. F. Smith, Injective modules and prime ideals, *Communications in Algebra* 9 (1981), 989–999.

M. H. Stone, The theory of representations of Boolean algebras, *Transactions of the American Mathematical Society* 40 (1936), 37–111.

H. Tachikawa, *Quasi-Frobenius rings and generalizations, QF-3 and QF-1 rings*, Lecture Notes in Mathematics #351. Berlin: Springer–Verlag, 1973.

C.-T. Tsai, *Report on injective modules*, Queen's Papers in

Pure and Applied Mathematics #6. Kingston, Ontario: Queen's University, 1965.

P. Vámos, The dual of the notion of "finitely generated", *Journal of the London Mathematical Society* 43 (1968), 643–646.

—————, Ideals and modules testing injectivity, *Communications in Algebra* 11 (1983), 2495–2505.

C. P. Walker, Relative homological algebra and abelian groups, *Illinois Journal of Mathematics* 10 (1966), 186–209.

R. Ware, Endomorphism rings of projective modules, *Transactions of the American Mathematical Society* 155 (1971), 233–256.

D. J. A. Welsh, *Matroid theory*, London: Academic Press, 1976.

S. Wiegand, Galois theory of essential extensions of modules, *Canadian Journal of Mathematics* 24 (1972), 573–579.

W. J. Wong, Finitely generated modules over principal ideal domains, *American Mathematical Monthly* 69 (1962), 398–400.

O. Zariski & P. Samuel, *Commutative algebra* (2 vols.), New York: Van Nostrand, 1958–1960.

Index